THE

ROUTEMASTER

YEARS

IN CROYDON & DISTRICT

by

Graham Donaldson

FOREWORD

I am often asked why I am interested in buses, a question difficult to answer save to say that they have been a source of unending fascination since the age of about 3. In those far off days I lived with my parents and grandparents in South Norwood, but can recall the trolleybuses swishing up and down the High Street, and frequent requests to my mother, when out shopping, to 'go and watch the buses go backwards' in Station Road at Norwood Junction. Alas since 1959 and 1982, respectively, it has not been possible to witness either of these things.

Apart from three years away at university in 1969-72 I have always lived in the Croydon area and have thus watched the many changes that have taken place over the years. Contrary to popular belief until recently, I had never done any job connected with buses - I say recently because I did help with the reissue of elderly persons' travel permits early in 1994! - although I did work in shipping in the early '70s. Most of my working career was with the Property Services Agency prior to its sale to the private sector, since when I have done various temporary jobs.

Several people have helped with the compilation of this book, not least the PSV Circle whose monthly News Sheets occupy most of a filing cabinet in my spare room, and which have been an invaluable source of information. Thanks is also extended to members of the Norbury & South London Transport Club, and to the several other photographers who have contributed to cover the early years, or to fill gaps in my own collection. Photographs in this book are by me unless otherwise credited.

One must, of course, add the usual riders to a publication of this kind. It is inevitable that, however much research one does and however carefully it is checked, a few (hopefully!) errors will creep in. I will of course be pleased to receive any comments, corrections or additional information. Secondly, this book is in no way an official publication of London Transport, London Buses or any other operator mentioned herein, and they are not responsible for anything said in it.

Some of what is written, particularly in the later chapters, may be considered controversial. I make no apology for that, as I feel that it is sometimes necessary from the point of view of setting the context in which things happened. But it should be emphasised that the principal target is the politicians and their dogma (not only, but particularly the present incumbents of Government), who seem so often to undervalue the role played by buses in keeping our towns and cities moving.

This book is therefore dedicated to the busmen and women of Croydon & District, past and present, without whom there would, after all, be no story....

Graham Donaldson
South Croydon
September 1994

INTRODUCTION

To many people, the Routemaster was the last 'proper' London bus, the last of a line started by the famous London General 'Old Bill' B types in 1910. Developed to the highest standards by London Transport engineers in the 1950s and 60s, it has perhaps achieved a unique distinction in becoming a legend in its own time.

This book does not seek to be a definitive history of the Routemaster, which has been well documented elsewhere. It seeks to tell the story of their operation in an area of South London, Croydon and District. The area described is based on that used in the useful area timetable books that were published by London Transport up to the end of 1969, and for a while thereafter by London Country Bus Services. It excludes routes terminating at Crystal Palace, other than in passing references, since these only just enter what is now the London Borough of Croydon and are not deemed to be an integral part of the Borough's transport system.

When Routemasters first operated in Croydon in the summer of 1960, they came to a very different town from the one we see today. The highest building in the centre was the Town Hall clock tower, and the site of the Fairfield Halls was a temporary car park. Whitgift Trinity School and its playing fields occupied the Whitgift Centre site and houses, shops and pubs stood where now there are office blocks. There was no bus station at West Croydon, and terminating services used a number of street stands in Station Road and St Michael's Road.

All this was to change in the 1960s and 70s as new development dramatically changed the face of the town. Routemaster operations reached their peak in the late 1960s before the advent of one person operation began to concentrate them more and more onto busy trunk and Central London services. In the 1980s, Government policy increasingly required London Transport to cut costs to the point where an inevitable consequence was the complete elimination of crew operated buses from the suburbs.

This, then, is a look back over a period of some 30 years during which Croydon witnessed the last flowering of the old order in suburban road transport.....

ROUTEMASTER TYPE CODES

The following type codes are applicable to Routemasters, and are used throughout the text:

RM - Routemaster (standard 64 seat bus).

RML - Routemaster Lengthened (72 seat bus recognisable by extra half-bay in each side of vehicle).

RMC - Routemaster Coach (57 seater with platform doors).

RCL - Lengthened version of RMC (65 seater).

RMA - Routemaster Airport coach (forward entrance 56 seat coach, originally built for British European Airways).

RMF - Routemaster Forward entrance. 69 seater. Only one example built, never operated in normal service in London.

FRM - Front entrance Routemaster, with rear engine, again a one-off. 72 seater.

CONTENTS

Early days: RM 283 in August 1960, working the extended route 64 to Wimbledon Stadium in part replacement for the 630 trolleybus. The style of shopfronts serve as a reminder of just how long the Routemaster has been around! *(Pamlin Prints)*

1 - 'THIS IS THE NEW ROUTEMASTER'

On a snowy morning in February 1956, the first prototype Routemaster, RM 1, took to the road for its first day of public service on route 2, Golders Green-Crystal Palace. It had been built some two years earlier, appearing at the 1954 Commercial Motor show at Earl's Court; but only when LT engineers were completely satisfied with the vehicle was it to be allowed to carry passengers in normal service. Such was the desire to achieve a standard of excellence, which is hard to imagine in the days of off-the-shelf purchases.

Three more prototypes followed: RM 2, in green Country livery; RML 3 with Leyland, as opposed to AEC running units; and finally CRL 4, an unusual combination of Leyland units and Eastern Coachworks bodywork, for Green Line service. In later life the latter two were to be reclassified, RML 3 becoming plain RM 3 (the 'L' was then applied to the 'Long' version of the Routemaster) and the coach becoming RMC 4 to accord with the 68 production models. None of these three vehicles was to see service in the Croydon area at this stage, however.

The primary purpose of the new bus was to provide a replacement for the London trolleybus. The capital then possessed the largest such system anywhere in the world but, following the demise of trams in July 1952, and despite the delivery of a number of new trolleybuses between 1948 and 1953, the writing was on the wall for this form of transport.

Croydon had only two trolleybus routes. The 630 was London's longest such service, running all the way from West Croydon to Harlesden (College Park) via Mitcham, Tooting, Putney and Hammersmith. The other service, 654, ran from Crystal Palace to Sutton (Bushey Road) via South Norwood, West Croydon and Wallington and required a fleet of dedicated vehicles fitted with special coasting and run-back brakes as a precaution against running away on the steep gradient of Anerley Hill.

However, had one Adolf Hitler not embroiled the Western World in conflict, these would not have been the only trolleybus routes in the Borough. Prior to 1939, the intention had been that all of London's tramways would be replaced by trolleybuses, which would have made good use of the electrical distribution network laid down for the railed vehicles. Accordingly tram routes 16/18, Purley-Embankment via Westminster or Blackfriars loop, would have become trolleybus routes 516/518. Plans reached an advanced stage, the planning authority having agreed, somewhat reluctantly, to the erection of overhead wires across the two bridges, and along the Embankment. This requirement had not arisen with trams which used the conduit - effectively third rail - system in Central London, switching from or to the overhead system at Streatham Change Pit.

Local tram service 42, Thornton Heath High Street-Croydon, would have become trolleybus 542. There was no suitable terminus for this at Coombe Road, where trams simply reversed their direction of travel, and so the 542 would have continued to South Croydon (Red Deer), using a turning loop via Sanderstead Road and Bynes Road.

As things turned out, World War II brought the tram to trolleybus conversions to a halt and, in 1945, the weary tramcars were still plying routes 16, 18 and 42 as best they could. But the 150 RT type motorbuses supplied in the Autumn of 1939 (not forgetting RT 1 which had arrived earlier) had performed superbly during the years of conflict, and the decision was quickly taken that the remaining trams would give way to RTs as soon as practicable. In fact this was to take some time to achieve, as the existing motorbus fleet of STs, STLs, LTs and other assorted types was in desperate need of replacement before the tram conversions could be tackled.

Eventually, though, the inevitable end came for Croydon's trams in 1951, routes 16/18 giving way to bus 109 and the 42 to bus 190. The last trams in London as a whole ran in the following year. This effectively sealed the fate of the trolleybuses and as the 1950s progressed so the cost of erecting and maintaining the overhead equipment escalated; and the emergence of one-way and traffic management schemes only made things worse. Buses were more flexible, declared LT management, they could go anywhere and relied on 'cheap imported fuel'. If only their crystal ball had foretold the oil crisis of the 1970s.....

With the age of the trolleybus fleet fast approaching the quarter-century mark, including those on route 654, a target of early 1958 (subsequently modified to early 1959) was set to begin their replacement. But it had taken rather longer than anticipated to start the production run of Routemasters, and it was clear that the number available by then would nowhere near meet requirements. The idea was floated of concentrating all of the available buses onto one ex-trolleybus route - and the 654 was suggested - but in the event a different plan emerged. This involved using some of the considerable surplus of RT and RTL types which were lying idle in various locations around the Metropolis - their services no longer required as car ownership rose and bus usage fell, year by year. The situation had been made dramatically worse in the Spring of 1958 when road staff held an all-out strike for seven weeks to demand better wages and conditions. When the buses did eventually return, many of their passengers did not, having found alternative ways to travel or, in some cases, not travelling at all. So began the spiral decline of service cuts and declining patronage which has continued, more or less, ever since.

Accordingly, newly overhauled RTs on route 154 replaced the trolleys on the 654 with effect from 4 March 1959, but were extended on weekdays from Sutton to Morden Station where they afforded a useful interchange with the Northern Line. In addition route 157 which had previously run only between Morden Station and Carshalton (Wrythe Green) was extended through to Crystal Palace.

Allied to this a rather curious route extension took place in order to provide more capacity between West Croydon and Anerley. This involved the 64, previously Addington-West Croydon, which was extended on weekdays (except evenings) via the old 654 to the 'Robin Hood' and thence to Elmers End Garage. During Monday-Friday rush hours it was further projected to Eden Park Station. These extensions, however, proved very short-lived and were withdrawn with effect from 8 July 1959.

The day of reckoning for Croydon's other trolleybus service came on 20 July 1960 when it was replaced throughout by new bus 220. This time Routemasters were used, operating from Shepherds Bush Garage rather than the former Hammersmith trolleybus depot which was closed. Once again the 64 was called on to assist, this time gaining an extension to Tooting and Wimbledon Stadium. This qualified it for part RM-operation and RM 257/63/5/70/5/82/3 were allocated to Elmers End, having previously seen service at West Ham and then, rather oddly, been delicensed.

The choice of Elmers End as an operating base was strange considering that it was really nowhere near the line of route. To take up duty, buses normally ran via route 194 to Elmers End Green, 54 to Shirley Park and 130 to Addington and one's recollection, vague though it is, is that some journeys carried passengers.

Most of the new buses carried a board under the canopy declaring proudly that 'This is the New Routemaster - London's bus of the future.' They attracted some attention - possibly even jealousy - from Croydon-based crews who were still using RTs on the service, and would continue to do so for another four years...

Precursor of the Routemaster (1): Tram No 393 pauses at Bensham Lane, Thornton Heath, shortly before replacement of tram 42 by bus 190. Prewar plans had envisaged this becoming a trolleybus service. ***(Pamlin Prints)***

Precursor of the Routemaster (2): D2-type trolleybus No 461 on the long route 630 from West Croydon to Harlesden, shortly before replacement. *(Pamlin Prints)*

In place of trolleybuses (1): Route 154 replaced the 654 trolleybus in March 1959, using surplus RTs. Saunders bodied RT 181 stands on the forecourt of Morden Station. *(Peter Relf)*

In place of trolleybuses (2): Route 157 was extended to Crystal Palace to supplement the 154. Newly-overhauled RT 3264 passes under West Croydon turning circle soon after the takeover; trolleys on route 630 would continue to turn here until the summer of 1960. *(Alan B. Cross)*

In place of trolleybuses (3): RM 344 works the new route 220 near West Croydon on 25 August 1960. The trolleybus overhead has already been dismantled. *(Pamlin Prints)*

A novelty on route 220 on 8 November 1964 was the unpainted RM 664, which was intended to test how the aluminium surface would stand up to the daily rigours of London life. Perhaps fortunately, the experiment was deemed a failure and London's buses stayed red. *(Gerald Mead)*

Thornton Heath took over Elmers End's share of route 64 in May 1962. RM 1176 stands at the Addington terminus on 9 September 1971, three months before Routemasters gave way to one-man DMS types.

2 - NEW AND NOT SO NEW

Things were fairly uneventful for Croydon's Routemasters for the next three and and a half years. Elsewhere in London, trolleybus replacement marched on relentlessly and from Stage IV of the conversion programme, implemented in November 1959, the Routemaster came into its own. The final stage came into effect on 9 May 1962 when trolleybuses based on Fulwell and Isleworth depots fell victim to the new order. Originally it had been the intention to retain these routes until about 1969/70 so that the postwar Q1 class trolleybuses could work out their economic lives, but procurement of buyers in Spain for the vehicles meant that nearly all of them left British shores in 1960. They were replaced by pre-war models decanted from converted North London routes, though what their crews and passengers thought about this is best left unrecorded....

Electric street traction in the London area had therefore come to an end, and was not to be reconsidered seriously until the emergence of the Croydon Tramlink proposals in the late 1980s.

Associated changes on 9 May saw Elmers End's share of route 64 transferred to Thornton Heath, along with the Routemasters. This type of vehicle was destined not to work from Elmers End again until 1973.

RMs into the 1200 series had all been used up on trolleybus replacement duties, and attention was now turning to replacement of the huge RT family. Originally this process had been intended to start in the suburbs, and because the new buses seated 64 passengers as opposed to 56 on the RTs the theory was that economies could be made by running fewer buses. However, this met with fierce opposition from the busmens' union and, after several months of failure to reach agreement, LT began to introduce Routemasters in place of RTs on Central London services, this being done on a straightforward one-for-one basis.

A notable event for Croydon occurred on 31 July 1963 when West Croydon Bus Station was brought into use. As noted in the introduction, buses had previously used a number of street stands giving rise to congestion in the town centre, and the new facility brought all terminating buses together on one site. But the arrangement was seen as only temporary, and so no shelters were provided for intending passengers! Not surprisingly this produced a torrent of complaints and the situation was quickly rectified.

A spin-off from the appearance of Routemasters on Central London routes was that they were often scheduled to work other services at weekends. Obviously, with schedules at these times calling for fewer buses there was scope for spreading the new vehicles more widely; the first instance to affect the Croydon area was on route 68, where RMs were scheduled from Chalk Farm Garage on Sundays with effect from 10 December 1963.

In January 1964 the time finally came for route 64 to receive a full complement of RMs, and a batch of new ones numbered in the early 1900 series was delivered to Croydon Garage for the purpose. For the first time, they were also scheduled to work route 234 (Selsdon-Wallington) on Sundays.

Thornton Heath's Sunday share of the 64 was transferred to Croydon with effect from 29 January and as a result, RMs from the former garage began to appear on route 115 on this day. This long, circituous route started from Whyteleafe on weekdays, Croydon Airport on Sundays, and ran via Thornton Heath Pond, Streatham and Mitcham to terminate at Wallington.

Another 'Sundays only' RM working occurred on route 40 (Wanstead Station-Herne Hill, extended Sundays to Norwood Junction), which received an allocation from 5 April 1964. Based at Upton Park, these derived from the 15 which had previously been worked with RTW types.

A change planned for the late summer was complete conversion of the 130 group of routes to RM. This would be a major undertaking involving routes 130, 130A (Saturdays & Sundays), 130B (Mondays-Fridays) and 130 EXPRESS (Monday-Friday rush hours). Of these the 130B had only been introduced as part of the January changes, providing a useful new link between Purley Way and New Addington.

Unfortunately for Croydon, Routemasters being delivered at this time carried illuminated advertisement panels on the offside, and it was decreed that these should be used on Central London services. Accordingly new vehicles were delivered to Dalston and Hendon to work routes 9, 11 and 13 and a motley collection of existing RMs was transferred to Croydon, with bonnet numbers including 225, 254, 373, 434, 599, 609, 630 and then straggling on into the 1800s. By the standards of the time, some of these were in surprisingly poor condition. The conversion process began on 1 September and was completed by the end of the year.

A problem arose on route 130B, where buses in New Addington ran alternately via Gascoigne Road or King Henry's Drive to reach the Industrial Area. RTs had carried a reversible slipboard by the front passenger window on the lower deck - the site normally used to publicise places of interest - to indicate which routeing was to be followed. On Routemasters, the corresponding board was placed beside the canopy route number box and thus could not be changed very easily. The solution adopted was a curious one, involving a sort of 'lazy' route display which read:-

CROYDON
Shirley
NEW ADDINGTON VULCAN WAY

with the 'destination' display reading 'VIA...' as appropriate. Buses working to Thornton Heath during rush hours used the conventional type of route screen and, as a result, wrong or confusing displays abounded. Consequently the 'lazy' blinds fell out of use after about eighteen months, and the destinations 'NEW ADDINGTON VIA GASCOIGNE ROAD' or 'NEW ADDINGTON VIA KING HENRY'S DRIVE' were used instead.

Not surprisingly, the arrival of so large a number of Routemasters caused them to appear on several other services at the weekend. Now they could be found on route 234 on Saturdays and Sundays, whilst Sundays saw them on the 59 (Chipstead Valley-West Hampstead); 68 (South Croydon-Chalk Farm) and 197 (Caterham Valley-Norwood Junction). With Norwood RMs being scheduled to route 68 on Sundays with effect from 1 November, the whole of this service was worked with the modern vehicles on this day. The Norwood vehicles were of particular interest in that they were new ones being allocated to routes 2B, 3 and 137, ending with RM 2217 which was to be the last of the standard length buses.

From 6 December route 115 was renumbered 115A on Sundays, and diverted between Hackbridge and Wallington to run via Nightingale Road and Carshalton. This replaced the shuttle service on route 77B between Tooting and Wallington on this day.

The effects of the 1964 London Government Act made themselves felt on 1 April 1965, with the replacement of the London County Council, and an assortment of County Boroughs, Metropolitan Boroughs and Urban Districts by the Greater London Council and associated London Borough Councils. The new London Borough of Croydon absorbed the Urban District of Coulsdon & Purley, while most of Penge Urban District, which had been the smallest authority in London, passed to Bromley. Initially, the expanded Croydon area included most of Warlingham and Farleigh - but so passionate were the wishes of people to remain in Surrey that some of this area was later transferred back.

The end of standard RM production, in the Spring of 1965, coincided with a change to the livery, which had been red and cream, in one form or another, since the war. The cream relief band on RM 2210 onwards was replaced by flake grey, which appeared almost white, and this styling was applied to other Central buses as they were overhauled. The new livery was not entirely satisfactory, however, and towards the end of 1965 the relief colour was changed to a more definite mist grey which remained standard for the remaining years of London Transport in its old form.

One of Croydon's 'motley collection' of 1964 was photographed at West Croydon soon after introduction. RM 254 was actually one of the better vehicles, having had its first overhaul. It is now preserved, but with a later body.

Route 234 was allocated Routemasters at weekends in 1964. RM 937 is seen at Purley on 19 November 1966. *(Colin Stannard)*

Sundays only route 115A was scheduled for RM operation between 1965 and 1971. RM 361 stands at the old Croydon Airport terminus (more recently supplanted by the Water Palace) on 29 May 1967. *(Gerald Mead)*

3 - GREEN ONES - AND RED ONES !

There was great excitement among bus watchers in 1965 when it was announced that 100 Routemasters, of the longer RML type, were to join the Country fleet. In many ways this was not before time; the Country Department had not seen any new buses - apart from experimental vehicles and leaving aside Green Line coaches - since the end of RT production in 1954.

Among the routes selected for their introduction were the 409 (West Croydon-Forest Row) and 411 (West Croydon-Reigate via Godstone) where they were to replace RTs. The launch day was Sunday 3 October, a beautiful sunny autumn day on which the new vehicles made a big impression in their smart lincoln green livery. Godstone Garage, since closed, proved to be a hive of activity with RMLs being prepared for service and RTs being despatched for storage at Reigate. For many of the latter, their service days were not over as they would be redeployed as red buses to replace some of the less popular RTL types.

The biggest surprise, however, was the fact that some of the RMLs in Godstone Garage that day were in RED livery! They were clearly intended for service although none took to the road on the first day, Monday 4 October seeing their first turn of duty. They even carried notices declaring that 'GREEN ROVER TICKETS ARE AVAILABLE ON THIS BUS' ! Evidently, not enough green vehicles had been delivered in time and, rather than have a phased conversion, LT decided to make use of buses for which the Central Department had no imediate use. It was to be the one and only time - aside from one-off emergencies involving Green Line workings - that red Routemasters would operate on Country routes. They lasted until the early months of 1966 by which time a sufficient number of green vehicles was available to replace them.

With Godstone's other main route the 410 (Bromley-Godstone-Reigate) also converting to RML all except two RTs were removed from strength, these remaining ostensibly for Green Line relief work on the 709 (Godstone-London-Chesham) and for special service 482 (Caterham-Smallfield Hospital) which ran on Thursdays and Sundays only.

The use of RMLs on the 410 was notable in that RTs had themselves worked it for less than a year, having replaced the lowbridge RLH class in November 1964. LT had tried for some time to persuade Surrey County Council to lower the road under Oxted station bridge so that normal double-deckers could be used, but to no avail. Eventually their patience ran out; buses were sent on a convoluted diversionary route to avoid the obstacle, and this was still in force when the RMLs took over. The diversion did, however achieve the desired result and, in the Spring of 1966, buses were able to resume their old routeing after the road surface had been lowered.

In addition to Godstone, East Grinstead and Reigate Garages each received a single Routemaster, RML 2306 (numerically the first green one) and 2308 being allocated respectively. Because of the complexity of Country bus scheduling both of these worked routes 409 and 411.

Green one: Gleaming new RML 2314 stands at West Croydon Bus Station on 9 October 1965. These vehicles made a big impression at the start. *(Pamlin Prints)*

Red one: RML 2296 pauses at Caterham on 10 October 1965, representing the only time that red Routemasters worked on Country services, apart from one-off emergencies. *(Gerald Mead)*

4 - THE SIXTIES ROLL ON

The year 1966 began gloomily with another round of service reductions on 23 January. Staff responded with an overtime ban which was to last until the end of February and, as a result, a number of routes was suspended so that a more reliable service could be run on those remaining. Route 234 was the only local service, on which RMs operated, to suffer from temporary withdrawal and was covered by a local coach operator on a short term basis.

Several Routemaster operated services featured in the programme of changes. Trolleybus replacement route 220 was withdrawn on Sundays between Tooting Station and West Croydon, leaving the 64 to cover this section alone. The latter route was once again reallocated on this day of the week, along with the 130A, so that Thornton Heath were in sole charge of both. Naturally, Thornton Heath's small RM allocation could nowhere near cope with all these workings and so both routes reverted to RT operation on Sundays - a complete reversal of the usual trend. The RMs meanwhile remained firmly entrenched on the 115A.

A minor change was the diversion of Sundays only route 59 to use Westminster Bridge instead of Lambeth Bridge.

During the Spring, Upton Park Garage received new RMLs to work the 15, and in consequence they began to appear on route 40 on Sundays.

May brought the allocation of RM types to Brixton for routes 95 and 95A where they replaced the last operational RTWs, so eliminating the first RT sub-class from London's streets. From 15 May RMs were introduced to another Sundays only service, 57A, between South Croydon Garage and Victoria. Route 109 had also latterly been worked with RTWs from Brixton but these were replaced with RTs; this busy service was not to see any Routemasters for another ten years. This, incidentally, was a salient point when, in the 1980s, 'The Sunday Sport' published a particularly ludicrous story about a Routemaster being hijacked from the 109 in 1968 and driven to the South Pole. The story became even more absurd when the vehicle illustrated was seen to carry a version of LT livery which was only introduced in 1974!

Things were then quiet until the Autumn when, from 6 November, Norwood Garage began to allocate RMs to the 68 on Saturdays as well as Sundays. Croydon and Chalk Farm followed suit from 1 January 1967, and appearances on Mondays-Fridays also became commonplace.

New Year's Eve 1966 brought another major round of changes, and this time route 220 was completely withdrawn between Tooting and West Croydon, with the 64 strengthened to compensate. At the southern end of the area, service 234 was withdrawn altogether on Saturdays and Sundays, the Purley-Wallington section being left solely to the single-deck 234A. On Sundays, the Purley-Selsdon section was replaced by extending route 57A from South Croydon to Selsdon; at the same time, it was withdrawn between Stockwell Station and Victoria. By now its relation with the parent service 57 was extremely tenuous; it might more logically have been renumbered 159B, but this was never done.

Replacement of the Saturday 234 to Selsdon was achieved by diverting route 115 to run there rather than to Whyteleafe - becoming 115B in the process - but keeping RTs.

The most noteworthy change, however, occurred on the Green Line front. The fortunes of these routes were generally in decline as falling custom and traffic congestion took their toll on what had been a very stable network since the 1950s. Route 709 was now dramatically reduced, with just a few journeys retained during Monday-Friday rush hours and on Sundays, between Baker Street and Godstone only. On these remaining journeys, the RF coaches were supplanted by RCL class Routemasters, three being allocated to Godstone after becoming surplus elsewhere. The retention of a limited Sunday service was perhaps surprising, but hospitals and convalescent homes in the Caterham area still gave rise to a fair number of coach-borne visitors. In consequence of the change, Godstone now became a 100% Routemaster operated shed.

1966 will perhaps best be remembered as the year when London Transport published its infamous 'Reshaping' Report. This envisaged a huge expansion of one man operation with single-deck, flat-fare single-deckers replacing crewed double-deckers on many suburban services. Trunk routes would remain two-man operated for the time being, but the Report envisaged that everything would be omo by about 1980.

This Report effectively sounded the death-knell for the Routemaster - or so it was thought at the time. Orders already placed would take their numbers to 2760, but after that there would be no more; a sad reflection on the original plan to build about 5000. A solitary rear-engined example, FRM 1, had been constructed in an attempt to make the model more acceptable in the new order of things - but it was too late. By Government decree, future buses were to be of the 'off the peg' variety and this decision was to have profound consequences in the years ahead. But more will be heard of the FRM later in the story.

The first part of 1967 was uneventful but then, on 16 September, a new route 130C was launched. This broke new ground by being an *all day* Express service between New Addington (Salcot Crescent) and West Croydon via the then new Fieldway Estate. It was worked by Croydon Garage RMs on weekdays, but on Sundays operated as a normal stopping service, extended to Thornton Heath High Street in place of the 130A, which now became Saturdays only. The Sunday service, for the time being, inherited the Thornton Heath RT working which had been introduced with the January changes.

All this made the 130 group the most complex in the Central area. Buses on the 130C EXPRESS used the traditional - for this type of operation - white on blue blind display with route screen wording 'EXPRESS via Fieldway'. To aid recognition, 130 EXPRESS blinds were changed from plain 'EXPRESS' to 'EXPRESS via Addington Hotel' - 'The Cunningham' of more recent years.

Moves were now afoot to convert the whole 130 group to RML operation, using some of the last vehicles to be built. As a prelude, to make room at Croydon Garage for the longer buses, route 64's weekday allocation was due to be transferred to Thornton Heath on 16 September. However, because of union objections concerning the loss of overtime the change was delayed until about the beginning of November. A spin-off from this was that Thornton Heath now, at long last, had enough RMs to work routes 64, 130C *and* 115A on Sundays!

The first RMLs appeared in November and were christened 'Jumbo' buses by the local press. Route 130 was converted first, followed by the other variants and, although most of the vehicles were new - this time - two of the 1961 batch of RMLs, 885 and 891, also managed to intrude.

On 30 December 1967 more Green Line Routemasters appeared on the scene when the half-hourly RF operated service on the 708 (East Grinstead-London-Hemel Hempstead) was replaced by an hourly RMC operated one. The vehicles had become surplus as a result of route withdrawals, and were split between East Grinstead and Hemel Hempstead Garages. They brought four types of Routemaster (RM, RML, RMC and RCL) to the Croydon area for the first time.

Route 57A received Routemasters in 1966; they lasted until December 1970 after which RTs returned. RM 1296 pauses in Croydon on 24 May 1970. ***(Gerald Mead)***

A surprise move at the end of 1966 was the allocation of three RCLs to the truncated route 709; incredibly, their stay was to last for nearly ten years. RCL 2237 was caught in North End, Croydon, on one of the Sunday workings on 19 March 1972.

By contrast, RMCs lasted little more than a year on route 708. RMC 1509 was photographed on 18 January 1969, with just under a month to go before RFs returned once again. *(Colin Stannard)*

5 - LAST DAYS OF THE OLD ORDER

RML conversion of the 130 group was completed in March 1968, although Croydon never got hold of the very last Routemaster, RML 2760, which took off on a series of overseas tours after delivery. It finally entered service at Upton Park, and thus no doubt did venture to the Croydon area on the Sunday route 40.

As with their shorter predecessors, the RMLs also appeared on routes 59 (Sundays); 68 (Saturdays & Sundays) and 197 (Sundays) - the latter taking them out to Caterham where they met up with green ones on the 409/411. Technically, no standard length RMs were now based at Croydon Garage.

The remainder of the year was a quiet period for the area. In furtherance of the 'Reshaping' Plan, large numbers of AEC 'Merlin' single-deckers were arriving, and being placed in store at various locations around London. There were no immediate plans for their use, and, indeed because of protracted negotiations with the unions none saw any significant use until 7 September. On this day, the first area 'Reshaping' schemes were unleashed at Wood Green and Walthamstow, involving the replacement of several time-honoured routes with new flat-fare single-deck ones. This finally brought home the idea that the old order was indeed changing - though there can be few who thought it was for the better....

At the turn of the year, rumours did begin to circulate that 'Merlins' were to be introduced on flat-fare services replacing parts of the 12, 166 and 197 serving the Norwood, Addiscombe and Thornton Heath areas. However, trials soon showed that the long single-deckers could not negotiate sharp corners at some of the locations involved, so the plans came to nothing. Further attempts at single manning of Croydon's services would have to await the arrival of the shorter AEC 'Swift' vehicles.

On 15 February 1969, route 708 lost its RMCs after a little more than a year of their use, receiving RFs once again. No further Green Line work could be found for the displaced vehicles, and so they were transferred to Hatfield Garage for use as buses.

Another service received RMs on Sundays with effect from 22 June; this was the 75 based at Catford Garage, which on weekdays ran between West Croydon and Woolwich, but which latterly had been curtailed at Blackheath on Sundays.

On the debit side, 20 July saw the 57A cut back again from Selsdon to South Croydon Garage, a new omo service 234B now running between these points using an RF.

The developing new estate at Roundshaw had also received an omo RF service, numbered 233, with effect from 31 May. This ran over the new Croydon flyover when opened, although during the first few weeks a temporary routeing via the Flyover slip road, South End and Southbridge Road had to be used as the new facility was not quite ready for traffic. The 39-seater RFs quickly proved inadequate and so, from 22 November XA class Leyland Atlanteans, newly overhauled following previous use in Central London, were introduced instead. This was the first instance of double-deck one person operation anywhere in London, although from now on this mode of working was to spread rapidly.

However, even this was to pale into insignificance compared with events on the political front. The Labour Government of the day intended to create 'Conurbation Transport Authorities', subsequently renamed 'Passenger Transport Authorities' in all the major cities outside London, to coordinate all forms of public transport. Although no such authority was proposed for the capital, opinion was swinging in favour of giving control of London Transport to the Greater London Council, and this was brought into effect on 1 January 1970 under the terms of the Transport (London) Act 1969. However, the GLC had little interest in the Country Bus and Green Line services since they ran way beyond its boundary.

Accordingly, these were to be vested in a new subsidiary of the newly created National Bus Company, to be known as London Country Bus Services Ltd, with headquarters at Reigate.

RMLs appeared on the 130 group from September 1967. RML 2740 stands at Central Parade, New Addington, on 7 July 1970, while working route 130B.

Route 130C was the only all-day Express service in London to see Routemaster operation. RML 2741 is seen at West Croydon on 7 July 1968. *(Pamlin Prints)*

On Sundays the 130C became a stopping service. RM 189 was photographed in Thornton Heath on 14 April 1968, at a time when Red Rover tickets cost just 6/- (30p)! *(Gerald Mead)*

One of the 1961-built 'intruders', RML 891, was photographed in Croydon on 24 May 1970, performing a Sunday duty on route 197. *(Gerald Mead)*

Sundays only route 59 also gained the benefit of long Routemasters; RML 2758 proves the point at West Croydon on 20 September 1970.

RML 2747 works route 68 in North End, Croydon, on 19 January 1969.

Route 75 gained RMs on Sundays in 1969. RM 1203 was caught at Penge on 16 July 1972.
(Gerald Mead)

6 - UNDER NEW MANAGEMENT

The seventies dawned and the red and green buses went their separate ways. A situation which had remained unaltered since 1933 was radically changed, and things would never be the same again. Although the handing of control to the GLC made some sense on strategic planning grounds, and was followed by many years of heavy investment in LT, the seeds of conflict were sown as Central Government and the GLC were often of different political persuasions.

There was little outward sign of change straight away. Green buses received 'LONDON COUNTRY' fleetnames in gold Johnson face capitals, though the method of application varied between garages, some making a tidier job than others. The 'GREEN LINE' fleetname of late had made no reference to London Transport anyway, but 'bulls-eye' symbols were gradually painted out. At some sheds this was a particularly leisurely process, and some vehicles could still be seen displaying this feature in September !

From about June onwards, the new London Country symbol began to make its appearance, and was sited on the staircase panel of Routemasters. Generally known as the 'Flying Wheel' or 'Flying Polo', the device was supposed to represent the encirclement of London by the new company, and the furrows of fields beyond....

Red buses now eschewed the traditional underlined style of fleetname, and repainted and overhauled vehicles began to receive 'LONDON TRANSPORT' plain gold transfers of a similar style to the London Country ones. This was never officially extended to RT or RF types - but some did have the new devices applied locally.

The first RT to RM conversion of the new decade affected route 68, which became fully RM operated from 7 February 1970. As a result, the last RTs were removed from Norwood Garage.

More radical changes were to come on 18 April. The 130 EXPRESS and 130C services were withdrawn and replaced by new flat-fare routes C1-C4 using XA Atlanteans using a 'farebox' system - the princely sum of 1/6d (7 1/2p) being charged. Route C1 replaced the 130 EXPRESS service, but ran during the day on Mondays-Fridays and on Sundays, the routeing in Croydon being via Wellesley Road. On Saturdays it became the C3, operating via North End for the benefit of shoppers - what a pity this is no longer possible ! Similarly, the C2 supplanted the 130C on Mondays-Fridays (via Wellesley Road), becoming C4 on Saturdays (via North End); there was no Sunday service.

Also for the benefit of Fieldway residents, service130B was diverted over the previous 130C routeing in New Addington leaving a section of Headley Drive without any buses. But it was felt that, on balance, this would only result in a short walk for a small number of people. Route 130A did likewise on Saturdays, and was reintroduced on Sundays.

Routemasters gained more new ground by being allocated to the 190 (Old Coulsdon-Thornton Heath High Street) on Saturdays, from Croydon Garage; the Sunday working was from Thornton Heath using RTs.

XA class Leyland Atlanteans made one more conquest under this programme, replacing RTs on route 234 on Mondays-Fridays using conventional one-man operation without fareboxes. The Sunday service had aready converted to omo with the introduction of route 234B during the previous summer; so a curious anomaly now was retention of the crew operated 115B on Saturdays - but this was to cease in October.

The XAs were soon joined by the solitary FRM which was technically available to work routes 233 and 234, though in practice it only appeared on the former. Because of its high driving position, the vehicle was deemed unsuitable to receive a farebox and so never worked the 'C' services. But the FRM soon became a firm favourite, giving a smooth, quiet ride compared with the rattle-box XAs which quickly deteriorated following their recent overhaul !

A change to Sunday rosters under the 18 April changes affected route 40, which lost its Upton Park RMLs in favour of RMs from Poplar and Camberwell. The service was withdrawn between Wanstead and Poplar.

A major conversion on 13 June 1970 was of routes 159 and 159A to RM. The latter service was a Saturdays only variation which operated via Lisson Grove and Rossmore Road in the St John's Wood area, instead of via Park Road. It was only extended south of Streatham Garage, to Thornton Heath Clock Tower, before 0930 and after 2230 hours. On the other hand, the parent 159 had a long standing extension from Thornton Heath to South Croydon Garage on Saturdays, providing a useful shopping link for those living in the Parchmore Road/Green Lane area.

From 14 June, Streatham RMs joined the Croydon-based RMLs on Sundays only route 59. This service, together with the 57A, covered all the ground served by route 159 which did not then operate on Sundays.

31 October brought more major upheavals with the introduction of the first AEC Swift SMS types to the area. These employed conventional one man operation but aided by automatic fare collection (afc) equipment which was supposed to be used by passengers having the correct money. In practice it did not prove much of an aid; people quickly found the machines intimidating, and in any case they - and their associated barriers - often failed or malfunctioned. The new buses were employed on routes 166 and 166A from Croydon Garage, 194 from Elmers End and 289 from Thornton Heath. In all cases they replaced RTs although three new RT operations (119A, 119B and 197A) were created at the same time.

The Sundays only 166A replaced the time-honoured 59 to Chipstead Valley, the latter diverting instead to Old Coulsdon in place of route 190, which was withdrawn on Sundays. Also on this day, route 130 joined its partner 130A in being reallocated to Thornton Heath, thereby receiving RMs instead of RMLs. Route 234B was introduced on Saturdays with XAs, replacing the anomalous 115B which was cut back to Purley - becoming plain 115 once again.

After just four months of RM operation, route 159A was discontinued as part of these changes. The 159 was now entirely diverted via Lisson Grove and Rossmore Road in place of the Monday-Friday 59A, which became a much truncated service running between Charing Cross and Streatham (Telford Avenue) during rush hours only.

In the history of Croydon area Routemaster workings, the 159A therefore holds the record for the shortest lived....

Routemaster operation of the 159A lasted only four months. RM 1735 was photographed in Whitehall on 13 June 1970. *(Gerald Mead)*

The New Order - and What might have been....

(Top): XA 11 at West Croydon works the new C2 service on 7 July 1970.

(Bottom): Solitary rear-engined Routemaster FRM 1 prepares to depart on route 233 to Roundshaw - a low profile existence for such a splendid vehicle. Had the FRM gone into production, the subsequent history of London's buses might have been very different.

7 - THE DMS ERA DAWNS

Even before the break-up of the old order, consideration had been given to the use of one-man double-deckers on a wider scale. The new single-deckers, with their multi-standing accommodation, had not been well received by a travelling public who were used to getting a seat. This was nowhere more apparent than on local routes 166 and 194, where protests abounded concerning the replacement of RTs by SMS types.

In comparative trials between the XA Atlanteans and the XF class Daimler Fleetlines, the latter emerged as the favoured type and so laid the ground for the DMS class. By the end of the decade this would be almost as large (numerically) as the Routemaster series, though ironically the first examples would have gone for scrapping before the last were delivered....

The DMS made its debut in 2 January 1971 on routes 95 and 220, operated by Brixton and Shepherds Bush respectively. The latter had, of course, been one of Croydon's pioneer Routemaster services, but now reached no nearer than Tooting. Conversion of route 95 affected the local area inasmuch as it temporarily removed all RM types from Brixton, causing the Sunday 57A to revert to RT operation.

The new type of bus was initially christened 'The Londoner' by its operator, in an attempt to bestow some sort of charisma on a par with that associated with the Routemaster. But the new title didn't 'catch on' with a vehicle widely seen as box-like and lacking in any character. The angular style of bodywork was similar to that employed on contemporary Manchester buses of the period - but in their all-red livery the DMSs lacked the flair of their Northern counterparts.

The changes introduced on 2 January also involved SMS types replacing crew buses on routes 115 and 115A, so setting a trend in removing RMs from one of their Sunday strongholds.

27 March saw route 196, previously Tufnell Park-Norwood Junction, converted to RM working, but withdrawn between Tufnell Park and Euston. New SMS operated service 239 was introduced between Tufnell Park and Waterloo. The same day also saw SMSs replace the XA/FRM operation on the 233; in future, the FRM would see service on routes 234 and 234B.

RMLs on the Sunday 197 schedule were replaced by SMS types from 28 March, but the single-deckers were diverted to use Cherry Orchard Road instead of Dingwall Road and Lansdowne Road, since they could not negotiate the sharp turn from George Street into Dingwall Road. The service therefore became 197B on this day.

The first DMSs to appear in Croydon did so on 24 July 1971, when they began to operate route 50 between Stockwell Station and Croydon (Katherine Street) via Streatham, Melfort Road and Thornton Heath. Because of adverse reaction to the previous all-red livery, at least these carried a white relief band along either side.

More were to come on 4 December, when they replaced RMs on Croydon's other pioneer service, the 64. This conversion removed Routemasters completely from Thornton Heath Garage, causing both the 130 and 130A to revert to RT working on Sundays - a situation which, in the case of route 130, was to last for nearly five years.

Routemasters were transferred into Croydon at this time for the conversion of routes 190 (Mondays-Fridays) and 197 (Saturdays), which effectively left only the Monday-Friday 197 with RT operation from this base.

The DMS Fleetline seemed all-conquering in the early 70s. New DMS 120 was photographed in Katherine Street on 31 July 1971. Route 50 was extended to replace the 133 at its southern end, having at one time been more familiar to Addiscombe residents.

Thornton Heath Garage became responsible for the 130A when it was reintroduced on Sundays. RM 665 was at Thornton Heath on 3 May 1970. *(Gerald Mead)*

Removal of RMs from Thornton Heath Garage, with the arrival of DMSs, caused both the 130 and 130A to regain RTs on Sundays in 1971. RT 4256 represents the 'new' order at West Croydon on 19 March 1972.

Following conversion of route 233 to SMS operation, the FRM was transferred to the 234/234B. It is seen on the latter route at Selsdon on 10 July 1971.

XA 40 illustrates the later livery carried by the type, incorporating 'open bullseye' fleetnames similar to the FRM. It is seen at West Croydon on 10 July 1971.

8 - DIFFERENT SHADES OF GREEN

The year 1972 was only eight days old when one of the last conversions of the SMS era took place, with the introduction of route 12A between South Croydon Garage and Peckham. This replaced the corresponding section of route 12 which was now truncated at Norwood Junction. Whilst this did not directly impact upon local Routemaster workings, it did clear the way for the 12 to be converted to RM, as previously the low bridge in Lower Addiscombe Road - near Bingham Road Station - had prevented this. Being slightly lower in height, RTs were allowed to pass under it, but only just.

It was to be the Country Routemasters, which had generally led uneventful lives since passing to their new owner, which were to feature prominently in the story in 1972. On New Year's Day - not then a Bank Holiday - the RTs on route 414 (West Croydon-Reigate-Dorking-Horsham) gave way to RCL class Routemasters that had been displaced from East London route 721 by new RP coaches. The 405 (West Croydon-Redhill-Crawley) followed suit on 25 March using RCLs similarly displaced from routes 704 and 705. Initially the RCLs ran in Green Line livery - but Reigate, Dorking and Crawley Garages lost little time in substituting yellow relief for light green, and 'LONDON COUNTRY' fleetnames for 'GREEN LINE'. The vehicles also soon began to carry posters.

The RCLs saw no use on Sundays. Route 405 had been converted to omo on Sundays in February 1971, and the 414 in March 1971. A mixture of RF, MB and MBS types was used.

Yellow relief had become standard, in place of cream, from June 1970 onwards on repainted vehicles, but it was only now that the Routemaster fleet began to be affected. From February the RMLs commenced overhaul at Aldenham Works, and the first ones emerged in lincoln green and yellow.

On 19 February, Godstone Garage placed a batch of 11 new Daimler Fleetlines in service on route 410 in place of RMLs. These introduced another new livery, with a lighter shade of green replacing the traditional colour. Combined with yellow relief extended around the lower deck windows, the overall effect was very pleasant. The Fleetlines, a diverted Western Welsh order, were originally to have been coded DF but a change of plan saw them classified AF, the 'A' apparently denoting an omo vehicle (vide the Leyland Atlantean AN class).

In April it was announced that all future repaints would be into the new 'mid-green' livery, and overhauled RMLs began to exhibit this version.

From 11 June route 411 was converted to AF operation on Sundays, the 409 being withdrawn on this day.

September brought even more dramatic consequences for the livery of the Country fleet. By decree of the National Bus Company, all subsidiaries were now required to adopt standard liveries of either leaf green or poppy red, in both cases with a white relief band. Corporate style fleetnames, incorporating the NBC's 'Double-N' symbol, and grey fleet-numbers and transfers, were also to be standard. In many ways this fulfilled the worst fears of many observers about the excesses of corporate state management. It seemed that individual company identities, often with a long and proud history, were to be all but submerged. And the new colours were not particularly attractive, even when newly applied....

Be that as it may, overhauled and recertified RMLs now began to appear in the requisite scheme of leaf green with a white waistband and grey fleet-numbers. Godstone's RML 2333 was the first, though this was not quite standard as the stock numbers were not of the NBC prescribed style, adhering instead to London Country's previous practice. The short-lived London Country symbol was now rendered obselete, the standard 'Double N' taking its place. This was particularly notable at some garages, where the new symbol could be seen mounted on an illuminated panel in the shape of the old one !

The fleet now began to look decidedly untidy, as it appeared that no two buses were in the same livery. This was compounded by a decision that using Aldenham for all bus overhauls was proving too expensive, causing many Routemasters to receive local attention, recertification and (usually) a repaint. It was all a far cry from the high standards of earlier times.

On the red buses, 2 September saw route 12 converted to RM at weekends from Peckham and Walworth Garages.

The red Routemasters were not entirely immune from livery changes at this time, as from October onwards white replaced mist grey as the relief colour for repaint and overhaul candidates.

RCLs provided a much enhanced standard of comfort on the long Country routes. RCL 2231 stands outside Reigate Garage on 8 July 1974.

The first of its class to operate in the new National green livery, RCL 2256 was caught at Crawley Boulevard on 8 July 1974.

RML 2316 remained lincoln green throughout its Country bus career. It is seen here in Katherine Street, Croydon, on 21 May 1977 by which time partial opo was imminent for the 409/411.

By contrast, RML 2317 demonstrates the short-lived 'mid-green' and yellow scheme at West Croydon on 7 July 1973. Note the 'flying wheel' device on the staircase panel.

Route 12A replaced the southern end of the 12 using AEC Swift SMS types - though their legendary failure rate caused Leyland Nationals to oust them in 1977. SMS 259 was photographed in Addiscombe on 15 May 1977, shortly before this happened.

Originally intended for Western Welsh, the 11-strong AF class rang the changes with London Country. AF 9 was caught in North Street, Bromley on 20 May 1978 working route 410. This was their weekday haunt, but Sundays saw them on the 411.

9 - BUSES FOR EXPORT

As 1973 arrived, a surprise development was the decision to withdraw the XA class completely for export to Hong Kong. All 50 were to go, even including XA 46-8 which had passed to London Country on its formation.

The type had certainly been problematic. It had been hoped that they would perform better on quieter suburban routes than they had in Central London, but this was not so. The express work on the 'C' routes had proved particularly arduous, and it was not uncommon to see more than one hapless vehicle broken down on the ascent of Gravel Hill !

Accordingly, new DMSs began to take over from 20 January, the 234/234B being done first followed by the 'C' routes. However, because of delays in getting the new vehicles into service, the process was not completed until 30 March when XA 22 became the last of its class to operate.

FRM 1 was delicensed at this time, but was subsequently overhauled seeing further service at Potters Bar, and then from Stockwell Garage on the Round London Sightseeing Tour. It was finally retired in 1983 - not a bad innings for a one-off prototype - and transferred to LT Museum stock.

From 6 January, DMSs had also replaced SMSs on routes 166, 166A, 197B, 233 and 233A (Croydon Garage) and 194 (Elmers End), bringing SMS operation at the former to an end. The Swifts went mainly to the Ealing area for further one-man conversions.

Also from 6 January, the protracted conversion of route 12 to RM was at last begun, Walworth's allocation being the first of the four garages to be dealt with. Elmers End's turn came on 10 March, with Routemasters returning there for the first time since May 1962. The customary weekend workings saw them on the 54 (Sundays) and 194B (Saturdays).

Route 12's conversion was completed on 12 May, on which date the 130A was reallocated to Croydon on Sundays thereby regaining RMLs instead of RTs.

A notable landmark during 1973 was official recognition, confirming what observers had long suspected, that plans to complete RT withdrawals in 1975 could not now possibly be realised. The Merlin and Swift single-deckers were proving very difficult to maintain, and even the Fleetlines were starting to show signs of trouble. As a result, routine maintenance on Routemasters started to suffer as fleet engineers battled to keep the newer vehicles roadworthy. Coupled with a mounting national shortage of spare parts, this led to a number of RM types being delicensed - and it was a problem that was to grow very much worse before it improved.

London Country's situation was no better, particularly in view of the lack of works back-up facilities following the untying from Aldenham, and by the end of the year several of their Routemasters were out of service. December casualties included the lone RML 2306 at East Grinstead and RML 2319 at Godstone; both were subsequently heavily cannibalised - and never ran again so establishing records for short-lived careers. Interestingly, East Grinstead retained an RML roster until July 1977 but never again had a vehicle; in practice one was borrowed from Godstone.

The developing vehicle crisis led to widespread cancellations as the 'NBA' (No Bus Available) syndrome overtook staff shortage as the main cause of difficulties in this area. The Reshaping process was brought almost to a grinding halt, but the 194B succombed to DMS operation from 1 December 1973 so losing its Saturday RMs. Route 54 was converted to DMS on Sundays, but received a Saturday RM allocation from Elmers End.

On a general theme, 1973 was also notable as being the year in which London Transport finally overcame union objections to the principle of women drivers. In future, 'One person operation' (OPO) would replace 'One man operation' (OMO) in technical parlance !

Route 197, with its 'out-county' working to Caterham, had long been in the frame for opo conversion; indeed the Sunday service had gone over in 1971. For some time, LT had been trying to persuade Croydon Council to modify the tight corner at the junction of George Street and Dingwall Road to permit DMS working - but to no avail. Eventually, patience ran out and, from 5 January 1974 the route was permanently diverted via Cherry Orchard Road and converted to Fleetlines. The Sunday number 197B now became superfluous and the suffix was dropped - literally, in a way, as Croydon DMSs did not have the number '197' added to their blind sets, using the suffixed display with the 'B' painted out.

This conversion technically ended RT working from Croydon Garage, but because of the RM shortage not all were removed straight away. To balance Sunday opo workings with those during the week, route 130A was converted to DMS on this day.

From the same date route 196 was revised, being withdrawn between Herne Hill and Euston and diverted instead to Brixton in place of the 2A which was withdrawn. This provided a link to the Victoria Line tube, which had opened through to Brixton in 1973. The route also gained a Sunday service for the first time, with the 40 being cut back to Norwood Garage.

From April 1974 the appearance of red Routemasters was changed quite dramatically with the introduction of a solid white roundel device in place of the gold fleetname, coupled with white fleet-numbers of a smaller size than those used previously. The effect was quite odd at first, but became familiar over the years. However, it typified the bland approach at that time which resulted in opo vehicles being painted all-over red; thankfully this practice, although tried out on RMs, was never extended to them.

There were to be no more changes during 1974. Because of the vehicle shortages, red RTs began to be repainted and recertified once again, and as the year closed they were still in charge of local routes 54, 75, 109, 119 and 119A/B On Sundays routes 57A (now truncated at Thornton Heath High Street) and 130 were also still scheduled for RTs. Appearances elsewhere were by no means uncommon.

(Top): RML 2306 was an early victim of the 1973 spare parts crisis, and its working life came to an end in December of that year. It is seen in happier times in Katherine Street on 7 July 1970 - the only representative of its type at the small East Grinstead Garage.

(Bottom): Route 194B was host to Routemasters on Saturdays for just nine months. Elmers End's RM 1858 was caught in George Street on 17 November 1973. *(Jim Blake)*

Bus operators in the 1970's, including London Transport, were gripped by overall advertising mania. RM 1676 was promoting 'English Apples & Pears' when photographed in January 1976. *(David Bowker)*

London Country joined in too, RMC 1490 touring several Routemaster-operating garages to promote 'London & Manchester Assurance'. An RT side blind was fitted at the front when photographed in the Spring of 1975. *(David Bowker)*

10 - HIRED BUSES TO THE RESCUE

Following considerable delay, Bromley and Catford Garages received RMs for route 47 from 26 January 1975, and this enabled the conversion of the 54 on Saturdays and 119B on Sundays, as well as part of the Saturday 119A.

New Fleetlines were now entering service as crew buses, classified DM, in many areas, to speed introduction of the type. From 2 March they completely replaced RMLs on the 130 group, and Croydon lost all vehicles of the latter type which were transferred to Stockwell for route 88. Needless to say, the unreliability of Fleetlines was now becoming legendary and in particular, a number of new vehicles quickly suffered gearbox failure. Substitutions by RMs on the 130 routes remained commonplace until opo conversion occurred in 1982, the 130B also retaining scheduled cross-workings from the 68 until that time.

Unlike the RMLs, however, the DMs were not used on routes 59 or 68 at weekends, and RMs took over these workings.

Starting on 22 March, route 54's Saturday allocation from Catford was converted to RM operation; the Elmers End share was, of course, already so.

Mention has been made of the continuing vehicle shortage affecting both London Transport and London Country. This was graphically illustrated in May 1975 when the latter purchased three secondhand buses from the Southdown company. These were Leyland Titan PD3 'Queen Mary' types registered 6949, 6933 and 6935 CD which were given fleet numbers LS (Leyland Southdown) 1-3 respectively. They were not repainted, however, remaining in apple green and cream but with NBC style London Country fleetnames, and entered service from Godstone Garage from July onwards as substitutes for RMLs. They made quite a strange sight, but were normally confined to peak hour workings.

Not to be outdone, London Transport announced in August that hired vehicles were to be used to alleviate the shortages, and three Leyland Titan PD3s - half-cab versions with traditional radiators this time - arrived in Croydon on hire from Southend Corporation. These were put to work as driver trainers, heralding the introduction of the type to public service on 22 September when they completely replaced RMs on route 190. The displaced Routemasters were quickly reallocated as much-needed spare cover elsewhere.

The full batch of hired buses comprised Southend nos 333-344 (CJN 433-44C), Massey bodied examples dating from 1965; and 345-7 (MHJ 345-7F), which had East Lancs bodies built in 1968. Normally only ten were based at Croydon at any one time; each Friday night one would return to Southend in exchange for a different vehicle, so that major maintenance could be done at the home station.

With their slow-revving engines and grinding manual gearboxes, the Leylands certainly met with a mixed reaction. All were actually younger vehicles than the RMs they replaced, but were not perceived as such even though they did help to keep the service going.

From 1 December, Sunday route 59 was finally converted to DM operation from Croydon Garage - but not the 68 which kept RMs. This was because the tight turn from Ernest Avenue into Knight's Hill, at West Norwood, was deemed too difficult for Fleetlines to negotiate.

The Southend Titans were destined not to last very long. LT was unable to reach agreement with the unions over further hirings which could have involved a variety of sources, and maintenance staff refused to deal with 'foreign' vehicles. Accordingly from 28 February 1976, route 190 became RM operated once again, all of the Southend buses returning whence they came, and bringing to an end a bold attempt to tackle the critical NBA situation. The saga did not enhance the reputation of LT in the eyes of the public, and little did anyone know then that, less than ten years later, London would start to see many different coloured buses running under contract with private operators....

During their short stay, the PD3s were not known to have worked any route other than the 190. It was fortuitous that Southend's indicator layout was not too dissimilar from that on the RM type, so that blinds from the latter could be used.

The problems over running of hired buses were not apparent on London Country, which was quick to seize the opportunity to avail itself of the Southend Titans. They spent the summer working from Harlow Garage.

On 3 April, the first RMC types appeared at Chelsham for routes 403 and 403 EXPRESS, and then from 1 May RMs from Bromley took over completely on the 119 and 119A. Later the same month, on 16 May, RCLs on the 709 finally gave way to Leyland National SNC coaches, so bringing to an end the scheduled used of Routemasters on Green Line service. That the 709 had lasted so long in its truncated form - nearly ten years - was remarkable since all other RM coach workings had ceased in 1972. The replacement service had a new timetable, and quickly proved less popular.

From 24 May 1976 a start was made on converting the Brixton share of route 109 to RM operation. This was a big task, the route then requiring a maximum of 62 buses shared almost equally between Brixton and Thornton Heath. The conversion process at the former was not completed until August.

Another outward sign of the continuing vehicle shortage was the allocation of RT 4154 to Croydon Garage in May, the first of the type for over two years. It normally worked route 190, but since RT blinds were no longer available it was fitted with some rather crude masking to the indicator boxes so that the smaller RM displays could be used.

1976 will be best remembered as the year which broke many long-standing meteorological records, with a long hot dry spell lasting from June right through to the August Bank Holiday when, characteristically, it rained heavily. On many days the temperature soared into the 90s Fahrenheit, and many rear-engined buses suffered cooling problems. This was most acute with SMSs for route 12A; by August, out of 32 buses allocated to Elmers End for this purpose, only six were serviceable. SMS 266 even suffered an engine fire outside the author's house at the time ! The RMs carried on regardless, but in rather smaller numbers due to the spares shortage.

By the late summer, however, London Country's position had improved enough to allow withdrawal of the ex-Southdown LS class, and RMLs resumed these duties from 4 September.

From 2 October, RMs began to replace RTs on Thornton Heath's share of route 109, as well as the Sunday 130 where RT working had even survived conversion of the weekday service to DM - a unique situation. With the completion of this process, routes 54 (Mondays-Fridays), 75 (Weekdays), and part of the 403 (Weekdays) became the only scheduled RT workings in the area. The writing was now on the wall for this type.

Route 75's turn came on 26 February 1977 with conversion to DMS operation, thus removing the Sunday RM workings. Latterly buses had terminated at Norwood Junction on this day, but since the terminus in Station Road involved reversing, the opo service was extended through to West Croydon.

From May onwards, RCL types began to join the RMCs on route 403. However, it was clear that available Routemasters and the surviving RTs would not be able to maintain the service, and so another hiring was arranged. This time the source was Maidstone Borough Council, which provided Leyland Atlanteans from the batch 27-32, 35, 37-9 and 43-6 (EKP 227-32C, JKE 335/7-9E, NKK 243F, OKM 145/6G).

Another surprise at this time was the recertification of three London Country RTs beginning with RT 1018 in May. This was followed by RT 3461 and 604 all of which received a repaint into National green livery and certificates of fitness for three years. They were the only RT types to see service in this livery. Route 403 therefore had a grand mixture of RCL, RMC, RT and hired Maidstone Atlantean types! Although all the RMCs were in National livery, some of the RCLs remained lincoln green and did so right to the end of their Country service.

From 9 July routes 409 and 411 were partially converted to opo using AN and AF types. Generally all through workings to East Grinstead and Reigate received these, RMLs only continuing to work between West Croydon and Godstone. An interesting side effect was that East Grinstead often used an XF, one of the experimental Fleetlines that had been contemporary with the XAs, but which had somerwhat longer working lives.

The tide of opo conversions was starting to spell the end for the green Routemasters, the new company having declared 100% opo as its ultimate aim. From 1 October 1977 route 414 was converted to Leyland National SNB operation, some of its RCLs moving to Chelsham in place of Maidstone Atlanteans. Route 405 followed on 26 November, but this time RCLs made a surprise move to Godstone, where they replaced some of the RMLs suffering certificate of fitness expiry. They used blinds which appeared to have been printed for the Maidstone Atlanteans, though these never, in fact, worked from there.

The record of 1977 would not be complete without recalling the silver Routemasters which were so painted to commemorate Her Majesty The Queen's jubilee year. Unusually the 25 buses, which were sponsored by advertisers, were given temporary fleet numbers SRM 1-25, and examples were to be seen at various times on routes 12, 59 and 159.

Silver Jubilee. (Top): One of the 25 RMs painted in the special livery was RM 1903, alias SRM 12. It is seen here on an early morning route 59 journey to Purley; the green space in the background is now Tesco's Superstore...

(Bottom): Standing on the forecourt of the old Streatham Garage SRM 19 (RM 1904) gleams in the early morning sunshine. *(Both photos: Ian King)*

To the rescue! (1): Former Southdown LS 1 stands at West Croydon on 19 August 1975. The NBC style London Country fleetnames sit rather oddly on traditional Southdown livery.

To the rescue!(2): Southend Massey-bodied Leyland PD3 No 335 stands at the Old Coulsdon terminus on 14 February 1976. RM blinds fitted these vehicles reasonably well, although the external winding handles were among the many novel features for London operation!

To the rescue! (3): Maidstone Borough Council No 31, a Leyland Atlantean, loads at Selsdon Road, South Croydon on 21 May 1977. Indicator blinds were made specially for these vehicles.

Route 54 was never officially RM on Mondays-Fridays, but in the weeks before opo conversion RT shortages made their use a necessity. RM 383 stands outside the Fairfield Halls on 17 April 1978.

11 - BUSPLAN

January 1978 should have seen the opo conversion of route 54 but, in the event, not enough DMSs could be scraped together and a stay of execution was granted until April. In practice, the number of RTs at Catford which were still fit for service was beginning to fall below the number scheduled, and so as the weeks passed more and more duties were covered by RMs.

Between 1 February and 3 March, the Brixton share of route 109 was converted to DM type crewed Fleetlines. These were of the 'quietened' B20 type although whether they really were quieter was a matter of some conjecture. They emitted a sound variously described as resembling a hair drier or a World War II air-raid siren; certainly, it was very different!

April saw the launch of the first stage of 'Busplan 78', which aimed to tailor the number of buses operated more closely to the resources available, and so improve reliability. On 22 April the RTs and RMs on route 54 finally gave way to DMSs, and so scheduled RT operation in the Borough came to an end. Now the only possible sightings of the type would be on the 119 or 119A (unscheduled) or the lone survivor, RT 604, on route 403. The Country example only lasted until July when it suffered engine failure, and was promptly delicensed. The RT type thus went out with a whimper in the erstwhile Country Area, compared with the unforgettable scenes that were to occur at Barking the following April, when red RT operation came to an end. RT 604 was eventually sold for preservation.

26 August saw the Thornton Heath 109 workings converted to DM, although this time using older examples culled from elsewhere, and Routemasters left this garage for the second time. At last, the Sunday 130 received DMs as well.

The RMs moved across to Catford and Bromley, where they replaced the last RTs in South London, on route 94. This ended the appearance of the type on the 119/119A; now, only Barking retained them for routes 62 and 87.

The next stage of 'Busplan' was implemented on 28 October and brought some far-reaching changes. A secondary objective of the scheme was to simplify the route structure and reduce the number of weekend variations. This spelled the end for the Sundays only 59 service, which was replaced by introducing routes 109, 159 and 190 on this day. The 109 service was between Brixton Station and South Croydon Garage only, using DMs from Thornton Heath. The other routes ran as normal with RMs, with the variation that the 159 ran via Westminster Bridge instead of Lambeth Bridge; it was a sign of the new order of things that it was *not* numbered 159A.

Another casualty was the 159's Saturday extension to South Croydon, which was abandoned with the rather lame excuse that it made the route too long. Residents in the Parchmore Road/Green Lane area would have to wait until 1988, with the introduction of route 250, before they again had the benefit of a through service to Croydon.

Route 119 lost its Monday-Friday rush hour projection along Purley Way to Thornton Heath Garage, but was extended all day via London Road and Brigstock Road to Thornton Heath High Street. The route was thus the same as on Saturdays, causing the 'A' suffix to be dropped. Curiously, though, the Sunday 119B continued to ply the tree-lined avenues of Shirley Way and Bridle Road whose weekday service (166) had converted to opo in 1970. Although no longer a scheduled RM working, it is worth recording that route 130A disappeared under this stage of 'Busplan', with the 130B gaining a Saturday & Sunday service between West Croydon and New Addington (Salcot Crescent).

Not to be outdone, London Country converted all workings on the 409 and 411 to opo from 28 October, using a mixture of old and new AN types. But this was not quite the end for Godstone's Routemasters as two - RML 2330 and 2344 - were retained as spare cover and continued to make appearances until July 1979. This proved the point that the distinction beween crew and opo working was considerably less rigid, on the Country buses, than on the red ones.

By the late summer, sufficient RMCs were available to enable the return of the Maidstone Atlanteans from Chelsham to their home base, and the last of the RCLs were removed by the end of the year.

'Busplan' brought an end to route 159's Saturday extension to South Croydon. RM 2174 heads through West Croydon Bus Station on 7 October 1978; the scene has since changed beyond recognition, as BT's Delta Point building now occupies this spot!

The number 119A disappeared under Busplan '78, the 119 effectively taking over its route. RM 348 looks to be in almost rural surroundings as it heads along Addiscombe Road on 7 October 1978.

New Leyland Atlanteans supplanted Routemasters on the 403; AN 199 was caught in Park Lane on 6 July 1979. In contrast to early versions of the model, these vehicles were to prove robust and reliable, with several being refurbished in 1993/94 for further service.

12 - LAST DAYS OF THE GREENS

Early in 1979, new AN types began to replace the RMCs on route 403. Initially, these were crew operated, but from 3 March the service was converted to opo. However, the 403 EXPRESS retained its Routemasters for a while longer, and scheduled crew journeys were introduced on route 453 (Caterham-on-the-Hill - Chelsham).

For London Transport, 1979 was another celebration year, this time to commemorate the 150th anniversary of the introduction of George Shillibeer's horse omnibus service between Paddington Green and Bank. Twelve RMs were painted in a version of Shillibeer's dark green, red and cream livery and were widely used during the year although none operated from local garages. Unlike the silver Routemasters, however, these retained their 'proper' fleetnumbers. More familiar in the Croydon area was the (numerically) last Fleetline, DM 2646, which was painted in Shillibeer livery at the behest of British Leyland, whose advertising it carried. It looked commendably smart, and worked route 109 from Brixton Garage.

The last RMLs finally left Godstone in July, 2344 transferring to Northfleet and 2330 joining the majority of its counterparts in returning to London Transport ownership. Of the 97 RML, 43 RCL and 69 RMC types that had passed to London Country virtually all were returned; the only exceptions were two RMLs which slipped through the net and were scrapped by Wombwell Diesels; and prototype RMC 4 which remained with the Reigate company for official preservation. However, another 17 RML, 2 RCL and 2 RMC vehicles only passed back to LT 'on paper'; being long-term non-runners that had been heavily cannibalised, they were sent for scrapping with reusable parts being returned to London.

London Transport was now pinning all its hopes on the Routemaster, in contrast with the disasters which had attended the introduction of Merlins, Swifts and later Fleetlines. Vehicles of the latter type began to depart in quantity following the withdrawal of the last RTs from service at Barking on 7 April. The future for the RM on Central London and trunk suburban services now seemed secure, it being publicly declared that they would last well into the 1990s. But, as usual, no-one could then tell what lay around the corner....

Green Line route 709 operated for the last time on 26 October. As a reminder of times past RMC 4, by now repainted in National green bus livery, was brought in to duplicate the final journey.

As the new decade arrived, operation of green Routemasters in the area was now confined to the 403 EXPRESS and 453 but many of the survivors were now looking very down at heel. This impression was not helped, in January 1980, by the arrival of an incredibly battered RML 2452 at Chelsham, later replaced by an equally unkempt RML 2445. Even at this late stage there were surprises to be sprung, RMLs having never operated from this shed before.

From 16 February 1980, crew operated ANs took over the remaining duties, and green Routemaster operation in the area ended after just over 14 years. Swanley's route 477 had the distinction of being the last on London Country as a whole, and on 1 March the company organised a farewell tour using RMC 1512 and RML 2446. This covered route 403 amongst others. Whilst this was a commendable enterprise, full marks were not due to whoever made the special blinds for these buses, declaring that it was a 'ROUTMASTER *(sic)* FAREWELL TOUR' !!

The 403 EXPRESS finally converted to opo on 1 September.

As far as the red buses were concerned, London Country's loss was their gain. Apart from the 'scrappers' mentioned earlier, all the acquired RMLs were returned to service following repaint or overhaul, and the RMCs were deployed on driver training duties although some members of the class were never actually used. The RCLs were, initially, only to be used for training but the critical Fleetline situation was eventually to see these, too, in use as red buses.

Realisation of the 'vintage' appeal of Routemasters, now that traditional buses were becoming rare outside London, led to the emergence of the 'Showbus' phenomenon. In various cases, enthusiastic members of staff adopted certain vehicles which were specially embellished, often with original features such as gold fleetnames, and maintained in rally condition. Fortunately, Croydon Garage had recently received RM 1000 from overhaul, and, with its prestigious fleetnumber and out-of-sequence registration 100 BXL, it was a natural for 'the treatment'. With cream relief band, old-style air intakes below the blind box and headlamps, gold transfers and restored off-side route number box, it looked as though it had travelled through a time-warp, and reminded observers of the subtle changes that had affected the appearance of the Routemasters during their long careers. It quickly became a familiar sight on the remaining crew duties from Croydon.

Elmers End followed suit with RM 64, and Streatham with RM 719. There were many others - although the accuracy of 'restoration' work varied, with some vehicles adopting livery styles which predated their construction! At first, management was happy to allow these variations from the standard livery for their publicity and goodwill value, but this was sadly not to continue....

All 69 RMCs received corporate NBC livery. RMC 1511 heads along Park Lane, Croydon with an Express working on 6 July 1979. These journeys kept Routemasters for a while longer than the stopping service.

RCL types also had a spell on route 403. RCL 2259, seen at South Croydon on 7 October1978, became the last lincoln green double-decker in service with London Country.

13 - CHEAP FARES: THE GLC v THE LAW LORDS

As 1981 arrived, Routemaster operation in the area had become fairly uncomplicated. The only routes now scheduled for their use were 68, 159, 190 and 196 (Daily); 119 (Weekdays) and 119B (Sundays). Appearances on routes 130 and 130B, in place of Fleetlines, were also commonplace.

From 5 April 1981, a flat-rate fare of 25p was introduced experimentally throughout the whole of the London suburban area. The aim was to simplify the fare structure and, although the arrangement was expensive for short-journey passengers, it did provide considerable savings for longer distances. The scheme was marketed as 'Fare Deal' and received extensive publicity. However, it did cast the first shadows of doubt over the continued use of Routemasters on suburban routes, since boarding delays on opo vehicles had been one of the reasons for their survival.

The Spring of 1981 also saw the return of a Labour GLC committed to a 25% fares reduction on LT bus and tube services, and work was immediately put in hand on ways of implementing this. The Election manifesto had also provided for the levying of a supplementary rate to pay for the reduction.

The only change to Routemaster workings during the year occurred on 15 August, when the Brixton share of route 109 reverted once again to RM working, the Fleetlines being required as part of complex vehicle moves associated with the opening of the new Westbourne Park Garage in West London.

The Cheap Fares policy, dubbed 'Fares Fair' was duly launched on 4 October amid a blaze of publicity. The whole GLC area was now split into zones although the suburban zone of the Fare Deal scheme was retained. Low graduated fares of 10p (short hop); 20p; 30p and 40p (maximum) were introduced according to the number of zones crossed.

It was immediately popular. Bus usage increased by some 25% and there was a perceptible reduction in traffic congestion; at long last, it seemed that the authorities had had the courage to do what had been urged on them for years. Fortunately, staff and vehicle shortages were now largely a thing of the past. Wage increases and the introduction of new Titan, Metrobus and Leyland National types enabled services to be more reliable than for many years, and to cope with the extra demand a number of unscheduled duplicates were run. The Routemasters, in particular, were now riding high following a blitz to return to service all RM, RML and RCL types that were capable of being so treated. The low seating capacity of the RMCs prevented their seeing further passenger carrying use at this stage, although the idea was considered.

But there were black clouds on the horizon. The Government would not allow the inclusion of British Rail services in the cheap fares scheme, and South London commuters were particularly unhappy about being faced with a supplementary rate followed by a BR fares increase. If only the Capitalcard /Travelcard scheme had been introduced at the same time, things might have been different....

As it was, Bromley Council under the leadership of Ald. Denis Barkway, decided to challenge the legality of the fares reduction in the High Court. The initial ruling, later supported by the Court of Appeal, was that the policy was legal - but only just. Then in December, the House of Lords ruled that in their view it was illegal, as it did not strike the right balance between the interests of fare- and rate-payers.

This decision confounded everyone, and at the time was probably the biggest calamity to hit London Transport since the strike of 1958. The immediate effect was a doubling of fares from 21 March 1982 which, overnight, reduced bus usage by considerably more than the increase previously recorded. Plans were now put in hand to reduce both tube and bus mileage, and to greatly increase the proportion of opo working on the latter.

The first 'normal' Routemaster withdrawals took place in August 1982. Apart from a handful of fire and accident victims, plus the non-runners acquired from London Country, no members of the class had previously been withdrawn and most were in excellent condition for their age. On 4 September, a massive programme of route changes was introduced and this took 200 Routemasters out of service, most of them never to run again.

Locally, these changes brought the complete withdrawal of the New Addington 'C' services, which were partly replaced by Express versions of routes 130 and 130B during rush hours only. The latter routes were converted to opo, mostly using the existing DM types which were suitably modified and reclassified D. The 130 was withdrawn between West Croydon and Streatham Garage, except on Sundays when it was extended to Streatham Hill (Telford Avenue).

Incidentally, 1982 had not been a good year for Express services, the 403 EXPRESS being withdrawn after 20 August.

To replace the Streatham Vale section of route 130, new RM operated route 60 was introduced between South Croydon (Garage or Swan & Sugar Loaf, depending on operational need) and Streatham Hill (Telford Avenue) via Greyhound Lane. Operated by Croydon Garage, this was the first use of the number 60 since the previous Colindale-Waterloo service, worked by Cricklewood RTLs, had been withdrawn in September 1968.

Another gain for Routemasters was the Thornton Heath share of route 109, as enough were now surplus to enable removal of DM types from these duties. The latter were mostly withdrawn, joining nearly 400 of their counterparts in the Elephants' Graveyard of the old AEC Works at Southall. Travellers on the Great Western main line out of Paddington could now see very clearly the consequences of the Law Lords' decision....

The most contentious change under the 4 September programme affected route 196, which had previously managed to retain RMs because of the need to reverse at Norwood Junction. Now opo conversion was decreed using Leyland Titans from Camberwell - the first of the type to reach the area. The obvious solution to the reversing problem would have been to use the crescent formed by Grosvenor Road/Belgrave Road, as indeed had been done in the case of short-working 157's when that route became opo in 1973. But objections from a few local residents were allowed to hold sway, and a convoluted diversion to the Clifford Road terminus (on the south side of the line) was adopted instead. This meant buses running via High Street, Sunny Bank, Manor Road, Portland Road and Doyle Road into Clifford Road, returning via Manor Road etc; this added nearly 1/4 mile to the route!

Regular travellers on the route, particularly commuters and shoppers, were rightly outraged at this diversion. The Clifford Road exit from Norwood Junction had only ever been open during rush hours, and at other times one had the choice of a lengthy walk along a footpath, or down a dark and unsavoury subway, to reach the new terminus. Bus management, however, seemed quite impervious to the criticism, reflecting a significant change for the worse under the new Government-inspired regime.

However, on a Saturday afternoon in November, fate was to wreak its revenge. Titan T 495, on leaving the Clifford Road terminus, failed to make the right turn into Manor Road and crashed head-on into the railway bridge, completely demolishing its top deck. Miraculously, no-one was travelling upstairs; had it been a Monday-Friday evening the bus would have been loaded with home-bound commuters, and the consequences could have been tragic indeed....

This unfortunate incident quickly achieved what protest had failed to deliver, and from 1 January 1983, buses were diverted to use the Grosvenor Road stand.

Route 60 was to see only seven months of crew operation, and from 23 April 1983 it was converted to DMS and reallocated to Thornton Heath.

The year 1983 should have been a year of celebration as it marked the fiftieth anniversary of the formation of London Transport. However the machinations of Government policy, and proposals to take control away from the GLC, cast a long shadow over events.

Nonetheless, a programme of garage and works open days was planned, and the appropriately numbered RM 1983 was painted in a special gold livery, doing the rounds of various garages according to the timing of their event. It visited Croydon in May and worked route 190; the open day itself proved very successful and involved several preserved vehicles. Amongst these were RT 1702 and Southend 341, complete with 190 blinds as in 1975/76.

A number of other vehicles donned special liveries as local iniatives. Chalk Farm Garage 'did up' RM 1933 into General livery, and this could occasionally be seen on route 68. Thornton Heath's DMS 1933 and Croydon's LS 194 also received this treatment, and very smart they looked. By a happy coincidence, the year also marked the centenary of Croydon's incorporation as a Borough, and so Fleetline D 2629 was painted into Croydon Corporation Tramways chocolate and cream livery, with gold block fleetnames and 'fleetnumber' 100. Earning the nickname of 'The Chocolate Box', 2629 had also been reupholstered with Titan style yellow seating moquette, and thus remained unique long after repaint back into red livery.

Golden Jubilee. RM 1983 loads outside Allders in May 1983 whilst working route 190 in the special livery. It looked quite smart in sunshine, but appeared rather dull otherwise. The vehicle saw subsequent service with both Kelvin Scottish and Carlislebus. *(Ian King)*

Showbus RM 1000 was caught in Station Road, West Croydon on 3 September 1982, the day before the infamous 'Law Lords' service cuts which brought the first mass withdrawals of Routemasters. Compare this photograph with the frontispiece....

Although converted to crew Fleetline operation in 1975, routes 130 and 130B continued to see RM workings until opo conversion in 1982. RM 1590 performs a stint in Katherine Street, Croydon, on 7 July 1979. Numerically, this had been one of the vehicles allocated to the route in 1964, but had subsequently served elsewhere.

RM 961 waits outside Norwood Junction station on 3 September 1982, the last day of Routemasters on the 196 and the last day that Station Road would be used as a terminus.

The Law Lords changes did not seem all bad news with the creation of new service 60 using Routemasters - though this was to prove short-lived. RM 982 was photographed in Croydon High Street soon after the route's introduction. *(Ian King)*

RMs returned to Thornton Heath Garage for the third time, as evidenced by RM 958 in North End, Croydon, on 27 August 1985. By then, route 109 had been curtailed at South Croydon.

14 - WRITING ON THE WALL

One positive outcome of the huge reductions which followed the Law Lords' decision was that enough resources became available for a much expanded night bus network. At the beginning of 1984, London's night buses - such as they were - still ran to a timetable directed at the historic needs of Fleet Street printing workers and early morning charladies. Few services ran on Saturday night/Sunday morning, when they might reasonably have expected to be busiest.

Route 109 had a night service of sorts, which had latterly been worked with Routemasters; the timetable was little changed from tramway days. All this was to change from 13 April, with the launch of a completely new night bus network bringing these facilities to parts of London and the suburbs which had never had them before. Locally, new route N78 began operating between Trafalgar Square and South Croydon (Swan & Sugar Loaf) on a half-hourly basis throughout the night, using D/DMS types from Brixton and Thornton Heath. The 109 night service was accordingly withdrawn, although some early morning journeys still continued.

The new services were an immediate success, and before long some journeys had to be duplicated on Friday and Saturday nights. It was the one bright spot in a scene of otherwise unremitting gloom.

In June 1984, the London Regional Transport Bill completed its passage through Parliament, and the Government took almost immediate action to place control of the capital's bus and tube services under the auspices of the Department of Transport. From 29 June, London Regional Transport assumed control reporting direct to the DTp; on the face of things, a not dissimilar arrangement from that which had applied before 1970. But it was made clear that LRT would not be the operating authority in the longer term; free standing businesses would be created to actually run the services. The Secretary of State also spoke in terms of bus routes being offered for competitive tender.

An amusing side-issue was the fact that many buses, at the time, carried posters exhorting Londoners to 'Keep GLC Working for London' , and immediate orders were given for their removal. However, the GLC had cunningly arranged a contract with an advertising agency which had some time left to run - so the Minister was, for once, forced to retreat.

The low state of morale began to be reflected in the standard of turnout of some of the buses. Routemasters, in particular, had always looked well cared for but now, advertising material was often removed without anything to take its place, leaving damaged and scratched paintwork for all to see. There was little point in producing in-house advertising in such a rapidly changing scenario.

Even the showbuses were not immune, management having given orders late in 1983 that they were to be returned to standard fleet livery on the grounds of causing confusion to the travelling public. Happily, however, these orders were ignored in many cases.

In the autumn it was announced that the B20 type Fleetlines, which under previous plans had been due for withdrawal by the end of 1985, would now be retained for a few more years as they would be able to facilitate the expansion of one person operation. Accordingly, between 19 and 27 October 1984 they replaced RMs on route 190, though retaining conductors for the time being. RMs did, however, continue to appear particularly on Saturdays, when the schedule called for two more buses than on Mondays-Fridays.

A second front of attack was mounted on route 119, where crew operated Titans began to replace Routemasters from 16 October; by the end of the month the RMs had all gone. The changeover on the Sunday 119B was particularly tidy; on 14 October the service was entirely RM, whereas from 21 October Titans were in sole charge.

January 1985 saw the launch of the long-awaited Capitalcard giving unlimited travel by BR train, Underground and bus services throughout Greater London - and on some buses beyond. This was a thwarted GLC initiative for which the Government tried to claim credit - perhaps not surprisingly in view of its success. Use of train and Underground services increased significantly; but the situation on the buses was rather stunted by the relentless tide of opo conversions, which now threatened to eliminate Routemasters from the outer suburbs.

On 2 February 1985, route 190 converted to opo. Local residents had staged a 'Save your local conductor' campaign in a last-ditch bid to prevent this; it was to no avail, as were many similar campaigns.

The number 59 made a comeback on a DMS driver-only service to replace the outer end of route 109. The intention had been for the latter to terminate in Central Croydon - but as agreement could not be reached with the planning authority on a suitable terminus, buses continued to run as far as South Croydon Garage (or Swan & Sugar Loaf in the evenings). The 'new' route 59 ran from Brixton Station to Purley, but was extended on Sundays to Chipstead Valley in place of the 166A, which was withdrawn; this temporarily restored the situation that had existed prior to October 1970! DMSs from Brixton and Thornton Heath Garages were used; many had spent a long period in store at Clapham, and still carried an obselete livery involving white upper deck window areas.

To provide some continuity of Sunday crew work, Thornton Heath gained a small share of route 68 on this day, buses generally working to or from the base in service. Accordingly, the running-in destinations 'THORNTON HEATH HIGH STREET & GARAGE' and 'WEST CROYDON & THORNTON HEATH GARAGE' began to appear when needed.

The withdrawal of the through service between Purley and Embankment severed a link that had existed since 1927, when Croydon Corporation Tramways joined forces with the erstwhile London County Council to provide this facility. No doubt few passengers had made the through trip in recent years, however.

Even route 159 was not immune from Fleetline encroachment, gaining a scheduled Saturday allocation of the type from Brixton.

On 1 April 1985, London Buses Ltd was formally constituted as an operating subsidiary of London Regional Transport. It had existed as a trading name since 1979 but now became the legal owner of the vehicles, a fact which was reflected in the name and address shown on the nearside panels.

27 April saw route 119 converted to opo and introduced on Sundays, the 119B being withdrawn. RMs had already gone, of course, but Bromley Garage managed to borrow RM 2211 to make a few journeys on the last day of crew working, Friday 26 April. The Sunday Shirley Way diversion was now covered by diverting route 194 - duly renumbered 194A - and this again harked back to the pre-1970 position, when route 194C had done the honours. That Shirley Way/Bridle Road had had an opo service on weekdays, and a crew service on Sundays for a period of over 14 years must surely be a record in the annals of London bus route history.

Conversion of the 119 to opo was particularly ill-received by the public, many complaining that the journey from Bromley to Croydon could take anything up to 20 minutes longer. But London Buses were unrepentant; costs had to be cut, and that was that.

In July 1985, the shape of things to come emerged with the offering of opo route 127 for competitive tender. Then in September, all double-deckers began to be fitted with advertisement frames into which card-mounted advertisements could be mounted without having to be stuck to the vehicle. By this means, it was hoped to overcome the problem of paintwork damage when adverts were removed; but the frames certainly did not enhance the appearance of buses particularly Routemasters, and were not an unqualified success.

November brought a further blow to the maintenance of vehicles with the cessation of Aldenham Works overhauls. No longer would it be possible to see a pristine Routemaster, looking as good as new....

RMs remained on the 190 for a surprisingly long time for an outer suburban route. RM 812 stands at the Old Coulsdon terminus on 6 September 1980, at a time when its future seemed secure.

The 1 in 7 gradient of Corkscrew Hill provided a challenge. The driver has to engage first gear as RM 671 reaches the steepest part during the evening rush hour of 10 July 1984, carrying a full load.

RM 386 rounds the corner from Addington Road into Corkscrew Hill on 10 July 1984. A roundabout now occupies this site.

Route 119B was to disappear upon opo conversion of the 119, having run on Sundays only for more than 14 years. RM 258 traverses Shirley Way, served on weekdays by the opo 166, on 23 October 1983.

Routes 119 and 196 were the only cases in the area where Routemasters were supplanted directly by Leyland Titans. T 305 was photographed at Hayes on 20 March 1993, by which time the Selkent company's name had made its appearance. The 'PAY CONDUCTOR' sign on the front was wishful thinking by this time....

London Country replaced Routemasters with Leyland Nationals on routes 405, 414 and 709. Many lasted into the 1990s such as SNB 349, here painted as an overall advert on 14 September 1991 against a background of the Delta Point block at West Croydon.

15 - BITTER HARVEST

The beginning of 1986 was particularly notable for bad news. The annual fares increase perhaps set the shape of things to come with the abolition of an old friend, the Red (Bus) Rover; in future a one-day all-zones bus pass would take its place.

Then it was announced that London Buses had been unsuccesful in retaining route 127 under the tendering procedures, and operation would pass to London Country on a date to be determined. By now, the 197 was also on offer to all comers.

Hard on the heels of this came the shock announcement that Elmers End Garage would close later in the year, because tendering losses had brought about over-capacity. There had been suggestions for some time that this garage might close in the longer term, but no-one had expected a decision so quickly.

From 1 February, London Buses finally got their way with regard to route 109, with the opening of a new stand in Park Street, Croydon, which buses reached by way of the flyover slip road and Park Lane. The terminus officially became 'CROYDON HIGH STREET' but, as most RMs were not equipped to show this 'CROYDON KATHERINE STREET' was more usual. Curiously, route 12A also began to use the new stand on Sundays instead of running to South Croydon as previously.

The Sunday 59 service to Chipstead Valley had not proved successful due to timekeeping difficulties, and so from 22 June the route was withdrawn south of Croydon (High Street) on this day. In its place, the 12A was extended to Chipstead Valley and renumbered 12B.

On 9 August route 197, along with new variations 197A and 197B, passed to London Country. Incidentally the latter service, which was the Sunday version, followed exactly the same routeing as the previous 197B operated between 1971 and 1974. As a result of these changes Croydon Garage had a surplus of Fleetlines; but they were retained on strength for the conversion of route 68 to opo, now planned for the autumn, and RMs were transferred away. Norwood began to allocate new Leyland Olympian L types to the route from the beginning of October, leaving Routemasters with a minor presence on what had seemed, until the recent past, to be one of their safe havens.

7 September saw London Country split asunder in readiness for privatisation, the Secretary of State having decreed this fate for several larger members of the National Bus Company. Most local services passed to London Country (South West), which continued to be based at Reigate. However, two of the other new subsidiaries also had interests in Croydon as route 483 (West Croydon-Sevenoaks) and part of the 726 (Windsor-Dartford) passed to London Country (South East); the remainder of the latter route became the responsibility of London Country (North West).

The closure of Elmers End Garage duly took place after traffic on 24 October, a DMS off route 194 being the last bus home at approximately 0050 hours on the morning of Saturday 25th. Route 12's RM allocation was transferred to Camberwell, and the service withdrawn between Penge (Pawleyne Arms) and Norwood Junction. All other workings passed to Croydon. For a few months, Elmers End remained in use as a store for new and withdrawn vehicles but, eventually, this ceased and the splendid postwar building - which had replaced the earlier one destroyed by a flying bomb during World War II - was demolished for housing development. The fact was not lost, on many people, that Conservative Government policies had thereby succeeded where even Hitler had failed....

Croydon became an all-opo garage with the conversion of route 68 from 25 October, and a batch of new L types became the official allocation for the service which lost its contributions from Chalk Farm and Thornton Heath. As there was still a need to find Sunday crew work at the latter, a share of route 2B was allocated; but as the days of RM working at this base were also numbered it was not deemed worthwhile to fit blinds for the service, and crewed Fleetlines were used for the few months that this working existed.

To counter criticism about increased journey times for London commuters as a result of the 68 conversion, a peak-hour Express service X68 was introduced on Mondays-Fridays between West Croydon and Russell Square using a batch of four Leyland Olympians with full coach seating. The service was 'with flow' only: to London in the mornings and to Croydon in the evenings. The coach Ls were based at Norwood Garage.

On a happy note, RM 1000, the Croydon showbus, was sold to the Garage Sports and Social Club now that its service career was at an end.

It had been hoped that at least the busy 109 would be spared opo conversion, but with the vice of financial stringency tightening ever more around London's transport system, this was not to be. However, at least this service did not suffer from being worked by doored vehicles in advance of conversion; indeed, in January 1987 RMLs began to arrive at Brixton ostensibly for route 159. Several made appearances on the 109 during the last few weeks of crew operation.

On 7 February 1987, DMSs from Thornton Heath and Metrobus M types from the newly-reopened Streatham Garage took over route 109, and this was preceded in the early hours by the presence of a considerable crowd at Thornton Heath to see RM 664 (numerically the one-time unpainted 'Silver Lady') arrive home at about 0020 hours on the final journey. This was the last RM operation to serve Central Croydon, and marked almost the end of an era. 'Driven into History' , reported the 'Croydon Advertiser' - but the rights and wrongs of this were hotly debated. It certainly seemed ironic that one could now travel by Routemaster, in the newly deregulated provinces, in places as far afield as Blackpool, Glasgow, Perth and Dundee.....but not in Croydon.

The new opo 109 marked a further break with tramway practice by not working in a loop around the Embankment, as had always previously been the case. Since a layover was now mandatory at both ends, all buses ran via Westminster Bridge to Trafalgar Square. To replace the Blackfriars section, route 59 was extended via the Elephant and Blackfriars Bridge to Farringdon Street; in practice the route was normally run as Purley-Brixton and Thornton Heath Garage-Farringdon Street.

Apart from routes just touching the fringes of the Borough at Crystal Palace, only the 159 now remained RM operated. Even this was converted to one person operation on Sundays, with effect from 12 July 1987, using T types from Camberwell and a mixture of L and M from Streatham.

Early in 1988, evidence of the continued retention of the B20 Fleetlines began to manifest itself with the fitting of new Iveco power units, in place of the original Leyland 'air raid sirens', to vehicles based at Croydon and Sutton Garages. These certainly gave a more lively performance, but at the expense of greater vibration.

On 14 May 1988, a new opo service 250 was introduced on weekdays between Brixton Station and Croydon (Katherine Street) via Green Lane, Parchmore Road, Brigstock Road and London Road, at last restoring the link that had been severed under 'Busplan' nearly ten years earlier. In consequence, route 159 was withdrawn between Streatham High Road (Green Lane) and Thornton Heath, except for Monday-Friday rush hours and Sundays.

RMLs came to route 109 at the eleventh hour, though they were really intended for the 159. RML 892 heads for Croydon in January 1987. ***(David Bowker)***

RM 911 crosses 'Goat House' Bridge at Norwood Junction on 11 October 1986. The sign points prophetically towards Penge, which would become route 12's southern terminus within a couple of weeks.

Time was running out both for RM 38 and Elmers End Garage when this photograph was taken on 27 September 1986. The RM subsequently became an immobile enquiry office for the Strathtay Scottish company, and was not finally scrapped until 1993. The garage, which replaced the previous one destroyed by a German flying-bomb, was demolished.

The 59 returned briefly to Chipstead Valley on Sundays in February 1985, but lasted only just over a year in this form. 'White top' DMS 2308 stands at the 'Midday Sun' terminus on 24 February 1985.

New express service X68, with its coach-seated Leyland Olympians, provided some compensation for the loss of Routemasters. L 169 passes Waterloo on 15 August 1991.

16 - THE NEW COMPANIES

The four-way split of London Country operations had provided something of a pointer as to what London Buses could expect in the light of Government privatisation proposals. It was eventually decided to split the company into 13 subsidiaries, and most of these were set up on a 'shadow' basis during the latter half of 1988. The exceptions were Westlink and London Coaches, who already enjoyed a considerable degree of autonomy.

A new 'LONDON BUSES' version of the famous roundel was devised, involving a red circle with yellow bar as distinct from the blue bar of the Underground company. In future, the plain roundel would denote London Transport as a whole. The new symbol was rather slow to make its appearance on buses at first, but gradually became a familiar sight, as did the new subsidiary names.

From 1 April 1989 the companies became responsible directly for most aspects of their operations, although ownership of the vehicles continued to rest with headquarters, buses being rented to the local operators as required. Locally most operations passed to South London Transport, which rather oddly took Tower Bridge as its symbol; but Selkent, London General and London Central also had a presence.

Routemaster service 159 became a joint operation between South London and London Central, and the former company used the route for an 'Inauguration Special' on Saturday 1 April, to mark the official start of the new era. Advance notice of the event was, however, greeted with some disbelief in view of the proposed date of operation! Two journeys on the 159 were extended from Streatham via the 59 to Purley; 197 to Caterham and 409 to Godstone Garage, recalling a working that had run in pre-war days. The vehicle used was, appropriately, RML 2305 which had been one of the red ones allocated to Godstone back in 1965. Special blinds were provided, but the vehicle demonstrated the only partial effectiveness of the aforementioned advert frames, as part of its offside poster was missing.

This working was the last occasion on which a Routemaster operated along North End, Croydon. Despite considerable public opposition, and warnings of delays to bus services, Croydon Council closed this road to all traffic in October 1989, and it has subsequently been pedestrianised.

The new companies seemed to have more success in winning contracts, and South London scored a considerable coup, in October, when it not only regained the 197 group but also route 403 which, of course, had always been 'green'. The new arrangements began on 28 October, although there was a partial involvement before this date due to the rundown of operations at London Country's Chelsham Garage.

On 20 July 1990, in temperatures comparable with the blistering summer of 1976, route 159's rush hour extension to Thornton Heath Clock Tower operated for the last time. Evidently the reduction to peak-hour status had, as is often the case, reduced traffic to the point where these journeys were no longer worthwhile. On this particular day, however, the run was probably accomplished more quickly than on Network South-East, where buckled rails caused heavy delays in the searing heat.

Effectively, therefore, Routemaster operation in the Borough of Croydon had come to an end, although routes 2B, 3 and 137 continued to touch the fringes when turning at Crystal Palace. Even these were not to last, however, the 137 being first to fall on 2 February 1991, when it was cut back to Streatham Hill (Telford Avenue). New opo service 137A took over the southern end.

The famine of RM operations in the suburbs has led to a number of 'Specials' in connection with garage open days, launching of new services or simply as a novel way of raising money for charity appeals. One such operated on Friday 5 July 1991, when London Central's Camberwell Garage recreated the old 40 Sunday route between Wanstead Station and Norwood Junction. Showbus RM 9 was used; unfortunately it had by then been reregistered OYM 374A, so that its 'cherished' VLT 9 plate could be donated to an Olympian coach. It had, perhaps, become a sad reflection on the values of the age that many RM registration plates were now worth more than the vehicles themselves!!

On 10 August 1991, another last bastion fell with the curtailment of route 2B at Norwood Garage, opo service 2A replacing the Crystal Palace end. In addition, the 3 was reduced in frequency between Herne Hill and Crystal Palace and partly replaced by the 3A, using the inevitable driver-only buses.

Further ramifications of the tendering process were felt on Leap Year Day, 29 February 1992, when London Coaches began operating the former Green Line service 726 between Heathrow Airport and Dartford. Originally London Transport, who had latterly subsidised the route, had proposed its complete withdrawal; fortunately, an unexpectedly large volume of protest from individuals and local councils saved this useful link. A batch of ten Ikarus-bodied DAF coaches, DK 1-10, was obtained to work the new service which was suitably launched by the actor Stephen Lewis, alias Inspector 'Blakey' of the 1970's ITV comedy series 'On the Buses'. In addition former Green Line RCL 2260, by now in the London Coaches sightseeing tour fleet, worked as a duplicate between Heathrow and West Croydon on the first day. This recalled summer Sundays during the 1950s and early 60s when double-deck reliefs - originally RLHs and then, following modifications to Worcester Park bridge, RTs - had worked the 725 between Windsor and West Croydon.

A fortnight later, on 14 March, the Brixton-Purley journeys on route 59 were effectively renumbered 109 - but no, there was no reappearance of Routemasters! By a strange quirk, however, on the very day that this change took place, work started to demolish the former Purley tram depot which had seen a period of use by milk tankers since closure in 1951.

More cataclysmic, perhaps, was closure of the splendid new Streatham Garage, which had only reopened five years earlier after a major rebuild. This was brought about by the loss of routes 166, 194 and 194B by South London to Selkent. It certainly seemed amazing that a Government which continually preached about financial stringency was quite prepared to throw past investment to waste in this manner....

The earlier changes to route 3 were reversed from this date, the 3A being withdrawn in favour of a strengthened - and still RM operated - service on the 3. However, opo conversion of the whole route was foreshadowed for later in the year.

Accordingly, the last day of 1992 was the last day of Routemaster working on this service, although RM 14 did manage a special appearance on January 1 1993 when the normal service was, of course, due to be opo anyway. As if by way of partial compensation for loss of their conductor operated facility, passengers from 2 January had the benefit of brand new Optare Spectra SP class vehicles, which certainly looked very distinctive. But teething troubles, most unusually, led to a few RMs creeping back during the fortnight or so following conversion. Thus, the RM era for Croydon and District ended where it began way back in 1956 - at Crystal Palace.

Ironically, 2 January was also the date chosen for Croydon's last operational Fleetline, DMS 2438, to work a one-off special service 459 from Chipstead Valley via routes 166, 109, 59, 45 and 259 to Waltham Cross and Hammond Street - a marathon run of more than three hours each way. This marked the 22nd anniversary of the DMS type first entering service. No 2438 was finally delicensed after service on 20 January 1993 and, thus it can be said that the Fleetline outlasted the Routemaster in the area - but only just!

23 October saw the Norbury and South London Transport Club hold its first exhibition at Croydon's Fairfield Halls, and in this connection London General kindly arranged for its vintage liveried RM 1590 to work special services to and from the event. In the morning, it worked return trips to Shrublands and Thornton Heath as a 194B, then spent most of the day working as an X30 to and from New Addington. Finally, a return run to Norwood Junction took place as a 197A. This generated considerable interest, particularly among children in New Addington, some of whom had probably never travelled on a 'real' bus before....!

A special service also operated for Sunday 24 October on routes X30 and 403 using the more prosaic GLS 501, a refurbished Leyland National from the Red Arrow fleet.

The early months of 1994 saw Routemaster operations ebb still further away from Croydon with the conversion of route 2 (renumbered from 2B on 20 March 1993) to driver-only buses with effect from 29 January. RML 2653 brought the era to a close becoming the last of its type to run into Norwood Garage at around 2030 hours on Friday 28th.

Meanwhile, route 159 had been put out to tender as the continuing joint operation by London Central/South London was viewed with disfavour in the run-up to privatisation. South London was the successful bidder, but the section between Streatham Station and Streatham High Road (Green Lane) was to be withdrawn. In addition, special liveried RM types were to be used rather than refurbished RMLs.

The new arrangements came into force on 29 January but, as only two of the dedicated RMs were ready in time interim use was made of the RMLs released from route 2. These were ultimately destined for London General to work the 11. The special livery turned out to be a markedly different, involving large areas of cream and resembling traditional Brighton, Hove & District practice. A 'line diagramme' for route 159 was carried along each side and several of the vehicles carried 'strange' registration marks, their original VLT, WLT or reversed CLT ones having been transferred elsewhere.

A few weeks later, on 2 and 3 March, a green Routemaster most unexpectedly returned to the area with the operation by London & Country of its newly acquired RM 1183 on route 408 during the morning rush hours. This vehicle had previously been with Southend Transport, and came to London & Country to alleviate a vehicle shortage following Southend's acquisition by British Bus plc - the same holding company that had purchased the Reigate operator. It was painted in the contemporary L&C livery of two shades of green with a red band.

This, then, is where we leave the Routemaster story - but, as the last paragraph shows, there is no telling when the last instalment will be written....

End of a long hot day. RM 1003 pulls away from Thornton Heath Clock Tower on 20 July 1990, the last day of Routemaster working to this terminus.

The April 1 'wind up' that really happened! RML 2305, numerically one of the red vehicles allocated to Godstone in 1965, works one of the special 159 journeys that marked the inauguration of South London Transport on 1 April 1989. *(David Bowker)*

RM 9 recreates the former route 40 working at Norwood Junction on 5 July 1991. It was not possible to use the old Station Road stand which, perhaps symbolically, had become a used car lot by this time.

RCL 2260 marked the inauguration of London Coaches' revamped 726. It is seen here in Croydon Road, Wallington, on 29 February 1992. *(David Bowker)*

'General' liveried RM 1590 stands at Central Parade, New Addington, whilst working the special exhibition service on 23 October 1993. In the time of the original London General company, this site would have been fields!

At the time of writing, the nearest one can travel to Croydon by Routemaster is Streatham Station, though the forecourt of the closed Garage still acts as a terminus. RM 1361, reregistered VYJ 808, stands in front of the boarded-up building on 12 July 1994. The special livery looks smart, but perhaps more reminiscent of Brighton than London....

Going in the other direction, London & Country's 4109 (RM 1183) could be found on route 410 on Summer Sundays and Bank Holidays. It is seen leaving Redhill Bus Station on May Day Bank Holiday 1994.

Goodbye...! RM 664 - numerically the 'Silver Lady' of the early 60s, prepares to leave Embankment late on 6 February 1987 with the last Routemaster journey on route 109. For nearly 60 years trams, and then buses, had provided a direct service between Croydon and the Embankment. *(David Bowker)*

APPENDIX I

ROUTEMASTERS ALLOCATED NEW FOR CROYDON AREA ROUTES

7/60 **Shepherds Bush (S)**:
RM 313/23-51/3/5-62/5-72/77/80 *

1/64 **Croydon (TC)**:
RM 1900/1/3/4/6/7/10

10/65-1/66 **Godstone (GD)**:
Red RML 2279/80/7/8/93/5-2305 **
Green RML 2307/10-17/9/29-36/46-54

10/65 **East Grinstead (EG)**:
RML 2306

10/65 **Reigate (RG)**:
RML 2308

9/67-2/68 **Croydon (TC)**:
RML 2662-7, 2716-23/5-7/9-33/9-55/7-9

* These also worked route 268 (Acton-Clapham Junction)

** Red RMLs were progressively replaced by green ones between 11/65 & 1/66.

Note: Codes are given for the garages concerned. The derivaton of most is obvious, but Croydon is coded **TC** reflecting its pre-London Transport ownership by Thomas Tilling. Similarly, Bromley is **TB** whilst Catford is **TL** for **T**illing's **L**ewisham Garage.

APPENDIX II

LIST OF ROUTEMASTERS AS ORIGINALLY SUPPLIED

Fleet Nos	Registration Nos	
RM 1	SLT 56	
RM 2	SLT 57	
RML 3	SLT 58	(a)
CRL 4	SLT 59	(b)
RM 5-300	VLT 5-300	
RM 301-879	WLT 301-879	
RML 880-903	WLT 880-903	(c)
RM 904-999	WLT 904-999	
RM 1000	100 BXL	
RM 1001-1253	1-253 CLT	
RMF 1254	254 CLT	
RM 1255-1452	255-452 CLT	
RMC 1453-1520	453-520 CLT	
RM 1521-1600	521-600 CLT	
RM 1601-1865	601-865 DYE	
RM 1866-1999	ALD 866-999B	
RM 2000	ALM 200B	
RM 2001-2105	ALM 1-105B	
RM 2106-2217	CUV 106-217C	
RCL 2218-2260	CUV 218-260C	
RML 2261-2363	CUV 261-363C	) (d)
RML 2364-2598	JJD 364-598D	)
RML 2599-2657	NML 599-657E	
RML 2658-2760	SMK 658-760F	
FRM 1	KGY 4D	

(a) Reclassified RM 3 in 1961.
(b) Reclassified RMC 4 in 1961.
(c) Originally to have been coded ER 880-903 (Extended Routemaster).
(d) RML 2306-2355 and 2411-2460 were delivered in green livery.

In addition, the 65 BEA Airport Coach Routemasters were acquired between 1975 and 1979 and numbered RMA 1-65, but in 3 different batches. Original registrations were KGY 601-625D and NMY 626-665E, but the RMA numbers did not tally in most cases.

APPENDIX III

NOTE ON LONDON TRANSPORT (ALDENHAM WORKS) OVERHAUL PROCEDURES

In the text, it will be noted in places that the term 'numerically' is used to preface comparisons of the same Routemaster bus at different points in time.

This is because of the extensive component changing that took place when buses were overhauled at Aldenham Works. When a vehicle arrived at works it would effectively 'disappear' , and the next bus to be despatched would take the next available stock number. Although the Routemaster is described as 'chassisless' it actually consists of a body and a front and rear sub-frame. At the start of the overhaul process the body was removed, and the two sub-frames clamped together so that they could be driven like a chassis. Because bodies took longer in overhaul than the sub-frames, it was very rare for the same components to be re-united when leaving works.

Each RM body carried a number prefixed by the letter 'B', and this was painted inside the canopy fairing on the nearside of the bus; thus one could keep track of what was where! Production bodies were numbered B5-2760, and there were two extra bodies (B9986/7) which were built as spares but were eventually used on vehicles that had suffered fire damage.

Towards the end of Aldenham overhauls in 1985, some effort *was* made to re-unite original vehicles; thus, for example, the RM 6 that still runs on route 159 actually is the bus which left the factory 35 years ago!

To aid the overhaul process, a 'Works Float' was maintained of vehicles which were reduced to components and, effectively, ceased to exist. This could mean particular stock numbers being off the road for many years at a stretch - very frustrating for junior bus spotters.

Sadly Aldenham and the associated Chiswick Works are now history, though the former were immortalised in the Cliff Richard film 'Summer Holiday', part of which was shot on location there.

MAP OF LOCAL ROUTEMASTER OPERATIONS

64 — (64) RM operated route with terminal.

59 — (59) Ditto - certain days of the week or rush hours only.

Other bus routes, not RM operated at any time.

This map shows all routes that had scheduled Routemaster working at some time between July 1960 and July 1990. *Note that they were **not** all concurrent.* The use of black or red figures on the map is dictated by the number of days of the week on which the route was RM operated, rather than when it actually ran.

Special note: Route 119 ran as shown until 10/78; thereafter via the 119A routeing but with the suffix discontinued.

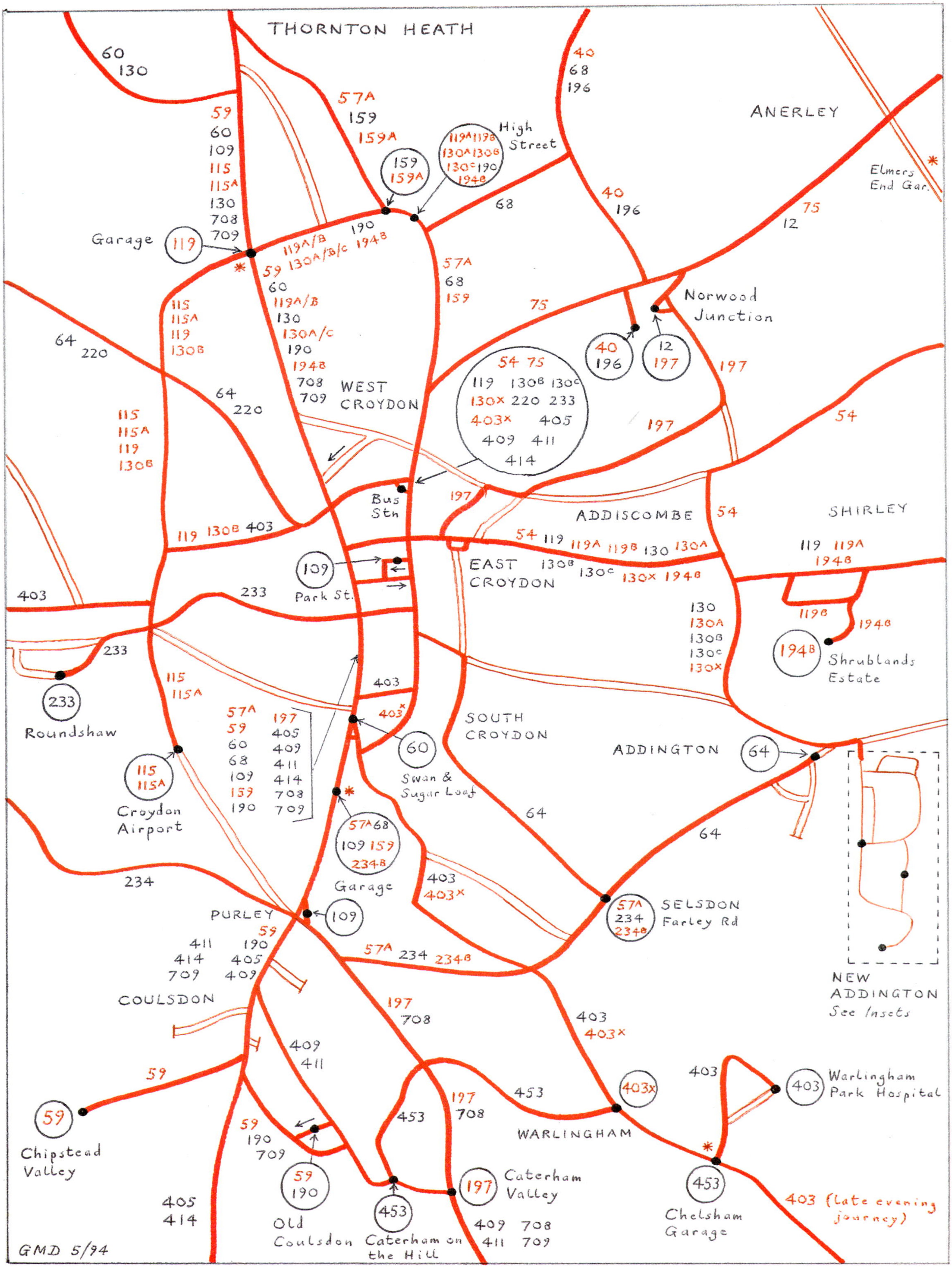

THORNTON HEATH
ANERLEY
Elmers End Gar.
High Street
Garage
Norwood Junction
WEST CROYDON
Bus Stn
Park St.
EAST CROYDON
ADDISCOMBE
SHIRLEY
Shrublands Estate
SOUTH CROYDON
Swan & Sugar Loaf
ADDINGTON
Roundshaw
Croydon Airport
Garage
SELSDON Farley Rd
PURLEY
COULSDON
NEW ADDINGTON See Insets
WARLINGHAM
Warlingham Park Hospital
Chipstead Valley
Old Coulsdon
Caterham on the Hill
Caterham Valley
Chelsham Garage
403 (late evening journey)
57A 159 159A
119A 119B 130A 130B 130C 190 194B
59 60 109 115 115A 130 708 709
60 130
40 68 196
75 12
57A 68 159
54 75 119 130B 130C 130X 220 233 403X 405 409 411 414
40 196
12 197
57A 59 60 68 109 159 190 197 405 409 411 414 708 709
57A 68 109 159 234B
57A 234 234B
59 190 411 414 709 405 409
197 708
409 411
453
59 190 709
405 414
409 708 411 709
GMD 5/94